HYBRID FIBER-OPTIC COAXIAL NETWORKS

How To Design, Build and Implement
An Enterprise-Wide, Broadband
Hybrid Fiber-Optic Coaxial (HFC) Network
that will carry Voice, Data and
Multi-channel, Bi-directional Video
and easily interconnect with standard,
carrier-provided ATM and SONET

by ERNEST TUNMANN

CRC Press
Taylor & Francis Group
Boca Raton London New York

CRC Press is an imprint of the
Taylor & Francis Group, an **informa** business

CRC Press
Taylor & Francis Group
6000 Broken Sound Parkway NW, Suite 300
Boca Raton, FL 33487-2742

First issued in hardback 2017

ISBN 13: 978-1-138-41249-1 (hbk)
ISBN 13: 978-0-936648-69-9 (pbk)

Visit the Taylor & Francis Web site at
http://www.taylorandfrancis.com

and the CRC Press Web site at
http://www.crcpress.com

To my wife
Margret
and my daughters
Jeannine and Linda

Contents

Chapter 2
Spectrum Utilization

Chapter 3
Analog and Digital Video Transmission

Chapter 4
The Gateway and Operations Center

Chapter 5
The HFC Broadband Network Components and
Performance

Chapter 9
The HFC Broadband Network Design Process -
Inside-Plant Design   183

Chapter 10
The HFC Broadband Network Design Process -
Outside-Plant Design 231

Chapter 13
The HFC Proposal Specifications

Preface

The contents of this book are devoted to the design and implementation of high-capacity, bi-directional broadband voice/data and video networks for small and large enterprises. Detailed information is provided relative to the planning, design and implementation of Hybrid Fiber-optic/Coaxial (HFC) broadband networks, as they apply to schools, universities, hospitals, manufacturing establishments or any other enterprises, whether they are contained in a single building or extended through multiple campus areas.

There is a growing need for more bi-directional transmission capacity in any enterprise. This growing need is often expressed in the desire to increase the transfer rate of the data network. Within the next few years, the need for multichannel, bi-directional video transmissions will multiply within every enterprise because of the coming integration of video into the conduct of business and the process of learning. Whether this video transmission will distribute distance learning, videography, multimedia authoring, desktop videoconferencing or video telephony, it requires a broadband cable to offer flexibility and growth potential.

Hybrid Fiber/Coaxial (HFC) has become the platform of choice for both cable TV and telephone companies for distribution of video, voice and data in the outside world. The reasons for the selection of HFC are economical and practical. HFC provides the highest bi-directional capacity at the lowest capital investment. HFC has become a favored broadband access platform because it is a moderate-cost, flexible transmission medium, capable of delivering heterogeneous traffic of varying bandwidth, modulation formats and latency characteristics in circuit-, channel- or packet-based formats. HFC also is scalable, allowing some capital investments to be deferred until customer demand emerges.

Full-motion video to NTSC specifications provides the enhanced picture quality required for diagnostic and educational video sequences. This high-quality video transmission can be provided today in a cost-effective manner by implementing a Hybrid Fiber/Coaxial (HFC) network throughout your enterprise.

The HFC broadband network is an open architecture. The HFC network will transmit any information, whether it is in analog or in digital form. Voice, high speed imaging, high speed data in T-1, FDDI or SONET formats, computer telephony, PCS or multi-channel video can reside on the HFC network simultaneously. All that is required is the selection of an RF frequency assignment for the service that you desire. The HFC network is the ultimate transmission platform. It empowers you as the network manager to select the frequency spectrum for the services that are particular to the needs of your enterprise.

This network design guide takes you step-by-step through the planning, the design and the implementation considerations of various HFC architectures. This network design guide also deals with enterprisewide NTSC video-conferencing between desktop workstations, with the establishment of distributed distance learning and telemedicine as well as with the accommodation of cable TV, video-on-demand (VOD), interactive television or any other video services that may be desired now or at any time in the future.

The design of HFC video networks is not rocket science. It is the author's opinion that anybody can implement an enterprisewide multichannel, bi-directional video network as long as the designer follows the step-by-step guidelines provided in this book.

Communication managers, LAN administrators, facility planners and anybody conscious of present and future full-motion video and high speed data requirements are encouraged to use this design guide to plan and implement an enterprisewide HFC network that will handle any future bandwidth requirements. At the same time, the HFC network will be able to interconnect with the outside world and the information superhighway using standard ATM and SONET technologies.

Your enterprise HFC network will be able to carry video, voice and data in analog and digital formats in exactly the same manner as the serving common carriers, giving your enterprise the competitive edge for any future equipment additions or upgrades well into the twenty-first century.

Your diligent work to establish the HFC network throughout your enterprise, while it may delay interim upgrade projects, will provide you with the capacity, flexibility and longevity to meet any transmission requirement of the future without the need for moves and changes. The HFC network will become the last and final infrastructure in your enterprise.

Chapter 1

Video Network Architectures

The Enterprise Environment

The transmission of video, high-speed data and video on the information superhighway requires to overcome long distances in a point-to-point or multipoint-to-multipoint environment. For these applications the packetization of all information and digitization for transmission on SONET survivable rings with ATM switches make sense and will provide the basis for the wideband switched network of the future.

Similarly to the historical development of telephony, the wideband transmission needs of private and public enterprises for long-distance connections will be occasional and consist of only a small percentage of the wideband transmission requirements needed for intracommunications within the enterprise. The PBX architecture with a small number of outgoing lines and large number of extensions is a good indicator of the difference between the large internal traffic load and the relatively small number of calls that are made to the outside world.

It is recognized that voice, data and video networks within enterprises require considerations different from those that appear logical for the information superhighway and different from the architectures proposed by cable companies for wideband services of the future.

While it is possible to digitize and packetize voice, data and video in the long-distance segment, voice, data and video networks within a building or a campus of an enterprise will remain separate. The main reasons for this separation are evolutionary.

Voice Transmission in Enterprise Networks

Telephone PBX installations are established and only need to be upgraded. At best, the replacement of a PBX system may require fiber-optic cables to node locations. Twisted pair copper cable then is used to outlet locations. This architecture reuses some of the existing copper cabling and is a proven and cost-effective solution for voice-system upgrades.

You may have installed an analog telephone system years ago and the development of data networks started with islands of automation. Perhaps now it is time to think about an upgrade to digital voice transmission throughout your facility. Such an upgrade may offer the opportunity to integrate data transmission into the new voice network.

If you take a futuristic perspective, you may decide to implement a T-I network, which lets you make easy configuration changes in the future. Whatever you do to upgrade the voice or voice/data network, it must be cost-effective and offer a migration path to ATM (Asynchronous Transfer Mode) switching for easy interconnection with public networks in the future.

If you take a more conservative perspective, you may decide not to change the existing telephone system, but to upgrade the data network with the goal in mind to fuse the two sometimes in the future.

Whatever your decision may be, it will accommodate the increased need for more and more telephone sets, computer telephone connections, moves and changes as well as additions to your corporate voice network.

PBX installations reflect the need for intracorporate communication by featuring ten and more times the number of extensions than outside lines. Traffic density of internal voice communication drives the decision to upgrade and the model to follow and complete the upgrade in an economical manner.

If internal voice traffic is many times the long-distance traffic requirements, does this rational apply to data and video as well?

Data Transmission in Enterprise Networks

High-speed data networks are frequently subject to upgrade requirements. Many existing enterprise networks feature isolated islands of automation where ethernet subsystems exist without any ability to interconnect or to reach the outside world. Depending on the size and area of the organization, FDDI and SONET ring architectures are recommended to integrate all data transmission throughout the enterprise and to accommodate a high-speed gateway traffic to the outside world. Single-mode fiber-optic cables between node locations can carry traffic at DS-3 speeds, while local or in-building traffic uses multi-mode fiber in conventional installation arrangements.

Other less far-sighted approaches may be T-I networking within the enterprise. Such a system can combine voice and data requirements but cannot accommodate multiple bi-directional video transmission requirements. T-I access to ATM is a very appealing migration path for the future.

Migration towards SONET (Synchronous Optical Network) is another option for increased speed in the data network. SONET is a form of TDM (Time Division Multiplex) framing on a transmission line that provides reference markers to tell the receiver how to interpret the bit stream. SONET can handle unframed transmissions such as T-I which, with a rate of I.5 Mbit/s, is a good match for the maximum wide area demand from a few LAN segments. Multiplexing T-I's produces DS-2 and DS-3. DS-3 at 45 Mbit/s fills an OC-I with a nominal bit rate of 5I.84 Mbit/s. Overhead makes the difference. OC-3 operates at a line rate of I55.52 Mbit/s.

An uncompressed video transmission to NTSC full-motion video specifications requires I5O Mbit/s when digitized. Compression technology at substantial cost can reduce this transmission requirement to OC-I or 45 Mbit/s or even to 6-9 Mbit/s using new MPEG-2 (Motion Picture Experts Group) compression specification. T-I compression of video may be the interim transmission choice for interconnection to remote locations via the public network. But, within the enterprise, the digitization of multichannel bi-directional video/audio would add a IOOO% load factor to the data network requirements and price itself right out of the market.

3

Video Transmission in Enterprise Networks

In my travels to consult with corporate entities that require the establishment of bi-directional video transmission throughout the enterprise, I am often told of unsolicited and solicited proposals by vendors and operating companies that are unresponsive to the communications needs of the enterprise. In many cases the use of fiber-optic cable is proclaimed to be the panacea for future growth of communications. T-I, ISDN, ATM and even SONET rings are offered to satisfy the video requirements.

In many cases these proposals are made without consideration of traffic requirements or to costs and longevity of the proposed installation. Multichannel bi-directional video, in digitized form, cannot simply be added to a data stream without overloading the facility and reducing the transfer rate of the data stream to a crawl.

If it is concluded that the internal video transmission requirements are greater than the video transmitted to and from enterprises through the public network, then video requires its own network.

Video networks will transmit video in analog form for a long time and until TV-sets with digital tuners are available at prices comparable to today's prices. Digitized and compressed video uses 9-45 Mbps for the transmission of full-motion video which, when large numbers of multiple video channels are required, would use up the capabilities of any SONET ring and ATM switch in the shortest time period. Cost considerations dictate that the multichannel, bi-directional video network requires its own cable system. A comparison of three different video network architectures is made in the following sections.

Gateway Considerations

The cable companies call the signal-origination location the "headend". In their view, that is the beginning of the forward delivery system. In the future, with two-way voice, data and video traffic, the headend will become the gateway to the outside world. The telephone companies locate the MDF (Main Distribution Frame) close to the enterprise PBX system. This is already a gateway concept, even though limited to voice and low-speed data. The data users within an enterprise look at high-speed data as an extension of the computer population and locate network management at the computer center. It is desirable to collocate the voice, data and video gateway locations. If that cannot be accomplished, than at least high-speed data and video gateways need to be in close proximity to permit digitization and packetization for communication with the outside world and over the information superhighway.

The Comparison Model

To compare fiber-optic, hybrid fiber-optic/coaxial and coaxial video systems, a model has been assumed. The model consists of headend circuit or channel-switching equipment and a distribution system for bi-directional video transmission to and from IOO outlets. A total of 2O-source inputs have been assumed. The IOO outlets are distributed over 6 buildings.

Cost considerations do not include source equipment, cable or TV-sets, but only compare transmission infrastructure equipment. The bi-directional capacity of each one of the architectures is enormous. Simultaneous transmissions of 8O forward and 3O return channels on one cable are state-of-the-art. Each channel is defined as a 6-MHz frequency slot that can carry full-motion NTSC video and stereo/audio.

The Fiber-Optic Delivery System

Since the early days of development of fiber-optic strands, the transmission of broadband services such as voice, data and video over fiber has been described as the coming revolution in telecommunication delivery systems. While this prognosis has come true for long-distance transmission systems, it will not become a factor in the subscriber loop or in the "last mile" considerations.

The reasons for this rejection of fiber in the "last mile" distribution are several:

> The cost of the transmission equipment that is required to translate electrical transmission to light and back again to electrical at the subscriber location.

> The inability of producing inexpensive coupling equipment that would permit the sending of a portion of the light to each subscriber.

> The inability of fiber to carry power to subscriber premises equipment.

> The need of dual fibers for send and receive transmission associated with a duplication of costs.

This means that fiber-optic cables directly to the subscriber's residence are not cost-effective, require an expensive equipment complement to be left at the subscriber's premises and require the subscriber to provide the power for the equipment.

The difficulties of using a dedicated fiber-optic delivery system in a campus area are shown in Fig. I-I. The model in Fig. I-I shows the similarities of a campus installation to the "last mile" subscriber loop. The headend, or video operating center, can be compared with a central office. Fiber cables emanate from the headend and are routed to each user, subscriber or outlet.

The problem with this application is the lack of distance. In a community, the single-mode fiber may be 20 km long. In a campus environment of an enterprise, the distance shrinks to I-5 km and the fiber-optic/electrical translation equipment costs become predominant.

Cost Considerations

Not considering the single-mode fiber-optic cabling required for a homerun installation to IOO outlets and not considering any source equipment, an analysis of the costs of the transmission infrastructure equipment shows a requirement for

IO fiber-optic transmitter/receiver units for forward transmission at a cost of about $5OO each and the total amount of $5,OOO.

IOO fiber-optic transmitter/receiver units for return transmission at a cost of about $5OO each and the total amount of $5O,OOO.

The matrix switch must be sized to permit the number of sources (2O) to be connected to IOO outlets. This requires a 20xIOO matrix switch with 2OOO video/audio crosspoints valued at about $6O,OOO.

For return transmission, the matrix switch has to be enlarged to accept the IOO inputs. This would add an additional IO,OOO crosspoints valued at $25O,OOO.

This transmission equipment related capital cost of close to $4OO,OOO does not include any headend equipment or TV-sets. All TV-sets for this homerun fiber-optic baseband system must be baseband TV monitors, which usually are more expensive than standard TV-sets.

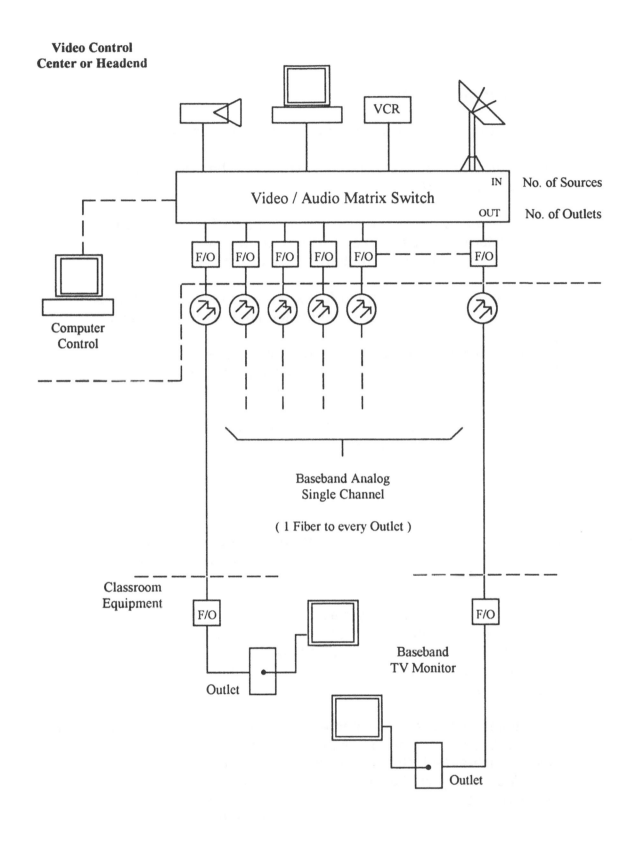

Fig. 1-1 The Fiber Optic Dedicated Delivery System

Technical Considerations

The number of electronic devices in this model is excessive.

The expensive matrix-switch modules have to be kept active, yet are rarely used.

The above analysis did not include the required transmission infrastructure, the fiber cable, which must be a single-mode type and which has to be routed directly to the classroom.

This type of installation cannot follow the data cable architecture, which requires type IIO crossconnect terminals at IDF locations.

Combined with the need for power and space for fiber-optic transmit and receive equipment in the classroom, this cable becomes an awkward nonstandard and costly proposition.

Moves and changes require the availability of fiber-optic spares at IDF locations on every floor.

Summary

The use of dedicated fiber-optic cables for transmission of video within a building or a campus area cannot be justified on the basis that fiber "is the way of the future".

The deficiencies such as too many electronic devices, difficult and nonstandard installation, complicated moves and changes as well as high costs of equipment that is not frequently utilized, precludes the utilization of dedicated fiber-optic cables, in the presence and in the future, for any "last mile" and especially for campus considerations.

Cable companies, RBOC's, AT&T and every other common carrier have stated similar concerns. Just imagine, a continuous fiber to your own home. The installation of such a dedicated transmission path would be impractical, wasteful and costly. The sharing of fiber-optic cables and transmission equipment for service to several customers is imperative to keep costs under control. In addition, this dedicated fiber could not even provide the ringing current for your telephone.

The Hybrid Fiber-Optic/Coaxial Delivery System (HFC)

Hybrid fiber-optic/coaxial delivery systems are the choice of all cable TV and telephone companies to deliver broadband voice, data and video transmission to the subscriber. Cable TV companies, in the past, have constructed coaxial cable systems throughout the license or franchise areas. These systems usually consist of single or multiple headends, a trunk and feeder (tree and branch) distribution plant and individual service drops to each subscriber. Since the number of amplifiers dictates the quality of the TV signal that is delivered, the availability of fiber to reduce the number of amplifiers was welcomed by the industry. Anytime a doubling of amplifiers in a coaxial system occurs, the C/N carrier-to-noise ratio is reduced by 3 dB. Multiple-channel TV pictures rely on a high (+43.O dB or better) carrier-to-noise ratio. Systems that have been built with more than (l6) amplifiers in cascade have problems meeting this FCC recommendation.

Another area of concern to the cable companies was the desire to increase channel capacity. This would require even more amplifiers in cascade because of the higher cable attenuation at higher frequencies. The desire to provide two-way data and voice services is another consideration that leads to use fiber-optic cables. Coaxial systems only have a very restricted return passband between 5-3O MHz. This band is also infested with a large body of interfering signals, which makes the implementation of voice and data a difficult task. The solution that can eliminate all of the above problems is a hybrid fiber-optic/coaxial architecture that, in addition, increases the availability of the system, a must requirement if telephony or data are to be transmitted.

The new cable-TV company system architecture then consists of fiber-optic cables leaving the headend in a star architecture and connecting to node points that can feed (lOO to 5OO) subscribers via coaxial distribution cable. In this manner, the number of amplifiers can be limited to a few in addition to the fiber/coaxial translator at the node point. This will greatly improve the picture quality as it increases the carrier-to-noise ratio and the reliability of the system to accommodate outage-free telephony, data and video-on-demand (VOD) services.

Can the fiber-optic/coaxial delivery system architecture be applied to a campus environment? The answer, of course, is yes, but not without a good analysis of all factors.
Fig. I-2 shows a model of a fiber-optic/coaxial delivery system utilizing fiber strands between the headend or video control center and the various building

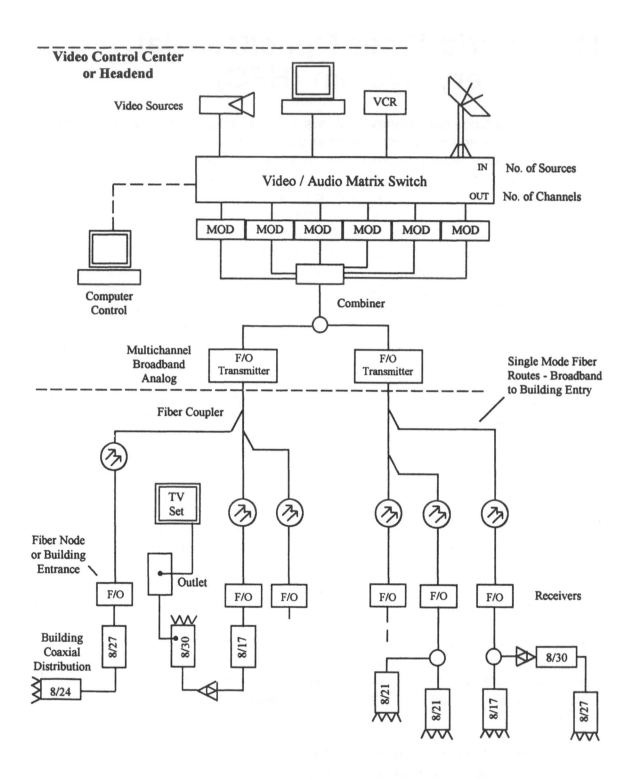

Fig. 1-2 The HFC Fiber Optic/Coaxial Delivery System

entry points. At the building entry location, the fiber transmission is translated to coaxial cable to accomplish the many splits to the numerous service locations.

It is recognized that the architecture, indicated in Fig. I-2, is subject to adaptation to the real world. Fiber to building entry locations only makes sense when other fibers are pulled in the outside-plant segment and economies of scale are achieved by this method. Our model is based on 100 outlets in 6 buildings, or an average of 16 outlets per building. This architecture may not be considered cost-effective.

Cost models of the industry show that a minimum of 125 users should be serviced from one fiber node. The more users can be served from one fiber node with coaxial cable, the lower the capital investment. 500 users have been targeted by both cable companies and common carriers as the upper limit for service from a fiber node. The reasons for this upper limit are plentiful -

> too many amplifiers
> reduced reliability
> traffic congestion for video-on-demand (VOD) services
> insufficient bandwidth for two-way voice and data services growth

Based on this rational, the model of Fig. I-2 should only contain one fiber node and feed the 6 buildings with coaxial cable. This would mean that only one fiber-optic strand would be required to connect the fiber node location with the video operations center. This, of course, would negate the fiber totally in favor of an all-coaxial system.

The deciding factor, again, is the distance between the video operations center and the 6 buildings as well as the distance between the buildings. If we assume, for a moment, that the buildings are 2 miles apart from each other, then the model of Fig. I-2 fits and commands fiber cable to be installed to all building entry locations.

Cost Considerations

Not considering the fiber-optic cable installation and any source equipment, an analysis of the costs of transmission infrastructure equipment shows:

> 20 modulators for a 20-channel transmission over the fiber network at $800 each, about $16,000

> 1 fiber-optic transmitter for 20 RF channels about $12,000

6 fiber-optic receivers for building entry points
at $2,000 each, about $12,000

coaxial cabling within 6 buildings and the service drops
are estimated at $30,000

the matrix switch can remain small for forward operation as the number
of outputs equals the number of channels.
A 20x20 video/audio matrix is estimated at about $12,000

In the return direction, it is possible to bundle the return fiber requirement
per building. The 100 transmit outlets require 20-channel fiber-optic
transmitters for each building entry point. This means 6 transmitters
at $15,000 each and 6 receivers at the headend at $2,000 each
for a total of $102,000

In addition, the matrix switch needs to be enlarged to accommodate
the number of simultaneous return transmissions. Assuming a
maximum of 12 simultaneous return transmissions, the matrix
switch has to be upgraded to a 32x20 configuration, which would
add an additional $8,000

The costs of the transmission equipment, required to establish a bi-directional
hybrid fiber-optic/coaxial system for 20 channels forward to 100 outlets and 12
channels of simultaneous return traffic, are estimated at about $190,000.

Technical Considerations

The number of electronic devices has been reduced

The capacity of the matrix switch has been reduced

The single-mode, fiber-optic cable terminals are at MDF locations
or at building entry points, which coincides with data network
installation standards

Moves and changes are easily accomplished by connecting
new drops to empty multitap ports of the coaxial system

Summary

The hybrid topology of a fiber-optic/coaxial video delivery system in a campus
environment has distinct advantages over the dedicated fiber-optic delivery system
in terms of costs, availability, protection from moves and changes and installation

matters. However, the costs are still high, especially when accommodating return transmission requirements. The hybrid topology is considered ideal for multi-campus and large single-campus systems. The use of expensive fiber-optic translation equipment for short distances (300 to 2000 ft.), as they may be typically found in campus environments, is recommended only if the number of coaxial amplifiers in cascade would exceed 2 trunkstations and 2 distribution amplifiers in a coaxial system.

The Coaxial Delivery System

Coaxial video delivery systems have been installed since the late 1950's to bring multichannel TV to over-the-horizon communities.

The channel capacity originally started with 5 channels. New-generation systems can handle up to 118 TV channels simultaneously. It goes to the credit of the cable TV industry that early vintage systems have been rebuilt at least two or three times. In other words, both hardware and service have been constantly changed and upgraded to keep pace with the ever-increasing demand for program choices and picture quality. All equipment, hardware and cable have matured and assure good reliability and improved longevity at reasonable prices.

A bi-directional coaxial cable system can be designed for an upper-frequency limit of 750 MHz. Using high-split crossover filtering, a single cable can provide a capacity of 88 forward channels and 30 return channels (see Fig. I-3)

Our model fits into any campus area when the distances between the video operations center and the buildings as well as the distances between buildings are in a range of 1000 to 3000 ft.

Distances dictate the selection of HFC over dedicated coaxial systems and are one of the most important consideration in the design of a video network. More details on the subject "distances" within an enterprise campus are discussed in the following sections dealing with single and multiple buildings.

Cost Considerations

Not considering cable installation and costs of source equipment, the costs of the transmission infrastructure equipment are estimated below:

The matrix switch can be kept small. It can be sized for the number of sources and return transmissions for inputs and the number of forward channels for outputs. A 32x32 video/audio matrix switch is estimated at about $32,000.

Again, a quantity of 20 modulators for outbound traffic has been assumed at $800 each, for about $16,000.

The distribution system extends throughout the campus area and feeds the 6 buildings in the model. The equipment costs of the distribution system, using the large size coaxial cable, high-quality two-way trunk and distribution amplifiers (capable of the transmission of a 750 MHz passband) is estimated at $20,000.

The remaining in-building coaxial-cable installation and the 100 service drops are, again, estimated at $30,000.

The cost of the transmission equipment required to establish a bi-directional coaxial system with a channel capacity of 88 channels in the forward direction and 30 channel in the return direction is estimated at about $100,000.

Technical Considerations

The number of electronic components has been further reduced. There are only 2 amplifiers in cascade, which will provide for excellent carrier-to-noise ratio and performance reliability.

The capacity of the matrix switch has been sized to meet future requirements.

The installation follows the data network architecture with amplifiers at MDF building entry locations and multitaps at IDF locations in risers of multi-story buildings.

Service drops can be added and changed to minimize any wiring requirements caused by moves and changes.

Summary

The old and proven coaxial cable tree-and-branch architecture is the winning solution for a single building or the multi-building environment of a small

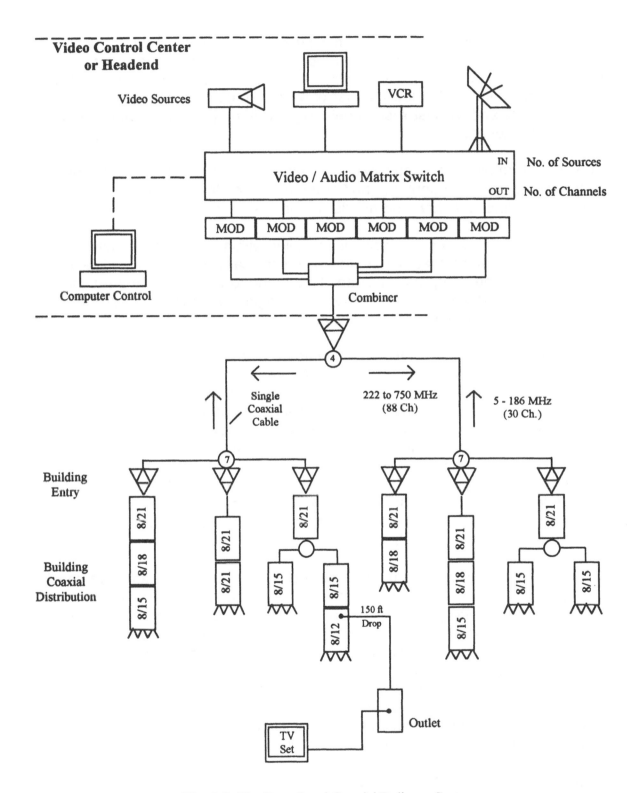

Fig. 1-3 The Broadband Coaxial Delivery System

15

campus of an enterprise. Coaxial cable systems utilize fairly inexpensive hardware components that have been upgraded many times over the past decades to provide reliable transmission in complex outdoor environments. In a campus environment, the low number of electronic devices will provide optimum picture quality and highest system reliability for many years and without the need of maintenance.

The cost of transmission equipment and hardware of a campuswide coaxial distribution system is about one-half of the capital outlay required for hybrid fiber/coaxial and less than one-quarter of the costs for a dedicated fiber system.

The transmission of video on fiber-optic cables within a small multi-building campus environment does not provide any technical performance or financial advantages to the user, now or at any time in the future, unless the number of cascaded amplifiers exceeds 4 or more and begins to compromise picture quality and the availability of the network.

Coaxial bi-directional transmission systems have to be engineered to meet performance requirements in excess of FCC requirements. It is recommended to use only cable television-proven components for amplifiers, passives and connectors. Drop cables must be quad-shielded or super-shielded to permit return transmissions without causing egress in high-level return transmissions. The length of the service drops shall be standardized. For instance, by cutting all drops to 150 ft., the outlet level can be controlled to be within +3.0 dBmV for all frequencies. A typical outlet specification is +8.0 dBmV +3.0 dBmV for receive levels and +54.0 dBmV +2.0 dBmV for transmit devices. Such a specification assures arrival of return transmission at the headend within a window of 3 dB for any location.

A multiplicity of two-way video channels is best carried on cost-effective coaxial cable, which can further the development of distance learning and campuswide NTSC teleconferencing. In conjunction with the high-speed data network, a cost-effective coaxial network promotes the establishment of desktop videoconferencing, with full-motion NTSC video-quality throughout the enterprise. The implementation of building and campuswide bi-directional video networks is essential to the development of multimedia teaching tools in corporations, schools, colleges and universities. It is also essential to the development of distributed telemedicine, continuous education and the learning of ever-changing medical procedures in hospital complexes.

Larger campuswide and multiple-campus video networks however require an HFC hybrid fiber-optic/coaxial architecture to assure optimum quality of performance. Fiber nodes can be selected to serve 150 to 500 outlets using coaxial cable. The number of service locations served by a fiber node is a function of density, the number of cascaded amplifiers and the distances between the facilities.

Dedicated fiber-optic video delivery systems do not offer any technical advantages for broadband networking and are not cost-effective. While the migration of high-speed data networks moves from FDDI to SONET rings with ATM switches, the most complex data network architecture cannot handle a large quantity of bi-directional video channels.

Recommendations for Single Buildings

Based on the foregoing model comparison, it becomes quite obvious that a high-quality, high-capacity, bi-directional, multi-channel video network within a single building should use broadband coaxial cable as transmission medium.

What are the limitations? Is the size of the building a factor in the decision to implement a coaxial system? How about the number of outlets or the location of the riser?

The following establishes average parameters for buildingwide broadband video networks.

Building Sizes

The size of a building can be expressed in square-feet per floor. Due to the 150 ft. service drop length limitation, the maximum square-footage per floor is about 18,000 ft.^{2}. This measurement assumes that electronic equipment can be located in the center of the building.

For a multi-story building, the riser distribution cable should be located in the center in order to accommodate all active and passive equipment at a centrally-located IDF.

Number of Floors

The number of floors that can be served by a coaxial video network is almost unlimited since the footage between floors is relatively small. By selecting the appropriate coaxial cable size in the riser, serial and parallel riser wiring allows quality video distribution in buildings with 1 to 32 stories.

Number of Outlets

The number of outlets signifies the total subscriber population in a building complex. Any number of service locations from I to 600 can be served by strategically placed distribution amplifiers. The design strategy must follow the main goal to minimize the number of cascaded amplifiers. The use of no more than 2 amplifiers in cascade is recommended.

Number of Outlets per Floor

There is no real limitation as to the number of outlets per floor. A quality distribution amplifier has the capacity to feed up to 60 service drops. An average number of 20 outlets per floor has been experienced in high-density environments.

Location of the Riser

In many buildings, the location of building entry locations and risers is predetermined by previous installation of voice and data services. Building entry locations are the cable entrance locations for connections with the outside world. The building entrance point commonly contains an MDF (Main Distribution Frame), which is used to terminate the cable that comes in from the outside world and a number of crossconnect panels that terminate in-building distribution wiring.

Larger multi-story buildings require a riser path from the MDF to the top floor of the building to accommodate voice and data cabling. The equipment closet at each floor level is sometimes called the IDF (Intermediate Distribution Frame). Again, these locations are used to terminate distribution wiring.

The coaxial broadband cabling mimics telephone and data network distribution practices. Passive devices such as splitters and multitaps are mounted in a convenient location on a wallboard. Service drops are connected to the multitaps and terminate at communication outlets at the service locations.

The location of the riser, with respect to the locations of the outlets, is an important consideration. Preferably, the riser is located in the center of the building so that equal-length service drops can emanate 150 ft. in each direction.

Risers located at the end of a building can only serve one-half of the square-footage of a building with a center riser. To compensate for such a deficiency, the designer may develop horizontal riser extensions between IDF locations and the geographic location that can support service drops of equal length.

Length of Service Drops

It is recommended that service drops are maintained at an equal length for all drops. Equal drop lengths will maintain similar attenuation values for all forward and return frequencies. This is an important consideration for maintaining tight outlet levels over the transmission frequency range. A ±3 dBmV window over the frequency band is desirable and can be assured by proper design. Longer drop lengths will increase the outlet-level window and harm the level control of return transmissions.

Number of Amplifiers

Single-building designs must accommodate the lowest number of cascaded amplifiers. While a small building can easily be served by one amplifier, large buildings, however, may require multiple amplifiers to satisfy the outlet population. It is important to configure the amplifiers in a parallel architecture and to minimize the number of amplifiers in cascade.

Maximum Amplifier Cascade

Distribution amplifiers have output levels in the range of +40.0 to +46.0 dBmV. In order to minimize the effects of noise and distortion, frequent reamplification must be avoided. The most important rule in designing high-capacity, bi-directional broadband transmission systems is the limitation of the number of amplifiers in cascade or in series. It is recommended to limit the amplifier cascade to a maximum of two.

Examples of Single Buildings served by Broadband Coaxial Cable

The table below indicates proven examples of four different buildings served by broadband coaxial cable. Outlet specification for levels of +8.0 ±3.0 dBmV were met for every outlet over the frequency band of 220 to 750 MHz.

	Small	Small	Large	Largest
Building size (ft.^{2})	9,000	18,000	86,000	280,000
Number of floors	1	2	4	16
Number of outlets	20	40	80	320
Av. outlets per floor	10	20	20	20
Location of the riser	side	center	dual	center
Length of service drops (ft.)	150	150	150	150
Number of amplifiers	1	1	2	4
Amplifiers in cascade	1	1	1	2

In conclusion, it is noted that broadband coaxial networks can be designed for any type and size of single buildings. The design can produce excellent video performance at all outlets within a 6 dBmV-level window and with a carrier-to-noise ratio of better than -53 dBmV. The cascading of amplifiers must be limited to two, which presents a real challenge to the designer.

Fig. I-4 shows the coaxial cable design for a I6-story building with 247 service drops. A total of 4 amplifiers have been used for building distribution. The 3 secondary distribution amplifiers in the third, ninth and thirteenth floor hubs are in a parallel configuration to the amplifier at the building entry location. Due to the architecture of the building, two categories of drop length were used. Short drops represent a uniform length of I2O ft., while long drops have been installed with a length of I8O ft.

The connection of the building to the trunk segment has been made by using a directional coupler. The trunk segment consists of a trunk amplifier, test tap and power supply. All of the trunk components and the building-entry distribution amplifier with test tap and directional coupler are located at the building-entry MDF closet.

Recommendations for Multiple Buildings

If single buildings can be wired economically using broadband coaxial cable, what are the implications for an enterprise campus area with multiple buildings? What method should be used to determine where the fiber-node locations shall be? How can I optimize the routing to obtain a cost-effective design? What are the limitations?

Obviously, the answers to these questions are directly related to the geography of the campus and the existing infrastructure. For instance, if you have space in the conduit plant and manholes are located in front of each building, you may want to bring fiber to each building. A 24-strand, single-mode fiber cable could be daisy-chained with breakouts for multiple fibers per building.

Another alternative would be to bring the 24-strand, single-mode fiber to the first building, translate to coaxial at the MDF of the first building, use 2 fibers (I for forward and I for return transmission) and save 22 spares for other services.

What are the limitations of any coaxial outside-plant segment?

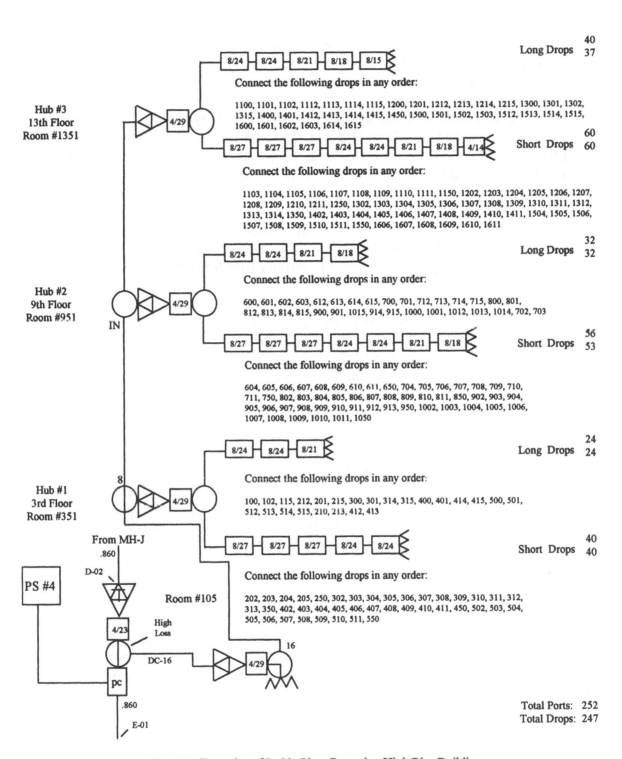

Long Drops 40 / 37

Hub #3
13th Floor
Room #1351

Boxes: 8/24 — 8/24 — 8/21 — 8/18 — 8/15

4/29

Connect the following drops in any order:

1100, 1101, 1102, 1112, 1113, 1114, 1115, 1200, 1201, 1212, 1213, 1214, 1215, 1300, 1301, 1302, 1315, 1400, 1401, 1412, 1413, 1414, 1415, 1450, 1500, 1501, 1502, 1503, 1512, 1513, 1514, 1515, 1600, 1601, 1602, 1603, 1614, 1615

Short Drops 60 / 60

Boxes: 8/27 — 8/27 — 8/27 — 8/24 — 8/24 — 8/21 — 8/18 — 4/14

Connect the following drops in any order:

1103, 1104, 1105, 1106, 1107, 1108, 1109, 1110, 1111, 1150, 1202, 1203, 1204, 1205, 1206, 1207, 1208, 1209, 1210, 1211, 1250, 1302, 1303, 1304, 1305, 1306, 1307, 1308, 1309, 1310, 1311, 1312, 1313, 1314, 1350, 1402, 1403, 1404, 1405, 1406, 1407, 1408, 1409, 1410, 1411, 1504, 1505, 1506, 1507, 1508, 1509, 1510, 1511, 1550, 1606, 1607, 1608, 1609, 1610, 1611

Long Drops 32 / 32

Hub #2
9th Floor
Room #951

IN

Boxes: 8/24 — 8/24 — 8/21 — 8/18

4/29

Connect the following drops in any order:

600, 601, 602, 603, 612, 613, 614, 615, 700, 701, 712, 713, 714, 715, 800, 801, 812, 813, 814, 815, 900, 901, 1015, 914, 915, 1000, 1001, 1012, 1013, 1014, 702, 703

Short Drops 56 / 53

Boxes: 8/27 — 8/27 — 8/27 — 8/24 — 8/24 — 8/21 — 8/18

Connect the following drops in any order:

604, 605, 606, 607, 608, 609, 610, 611, 650, 704, 705, 706, 707, 708, 709, 710, 711, 750, 802, 803, 804, 805, 806, 807, 808, 809, 810, 811, 850, 902, 903, 904, 905, 906, 907, 908, 909, 910, 911, 912, 913, 950, 1002, 1003, 1004, 1005, 1006, 1007, 1008, 1009, 1010, 1011, 1050

Long Drops 24 / 24

Hub #1
3rd Floor
Room #351

8

Boxes: 8/24 — 8/24 — 8/21

4/29

Connect the following drops in any order:

100, 102, 115, 212, 201, 215, 300, 301, 314, 315, 400, 401, 414, 415, 500, 501, 512, 513, 514, 515, 210, 213, 412, 413

Short Drops 40 / 40

Boxes: 8/27 — 8/27 — 8/27 — 8/24 — 8/24

From MH-J
.860

D-02

PS #4

Room #105

4/23

High Loss

DC-16

pc

.860

E-01

4/29

16

Connect the following drops in any order:

202, 203, 204, 205, 250, 302, 303, 304, 305, 306, 307, 308, 309, 310, 311, 312, 313, 350, 402, 403, 404, 405, 406, 407, 408, 409, 410, 411, 450, 502, 503, 504, 505, 506, 507, 508, 509, 510, 511, 550

Total Ports: 252
Total Drops: 247

Fig. 1-4 Examples of Inside Plant Records - High Rise Building

In our single-building design, we have determined that we will never have more than 2 distribution amplifiers in cascade. The recommended standard permits 2 trunk-amplifier spacings and 2 distribution amplifiers. This means that our outside-plant trunk segment can utilize 2 trunk amplifiers in cascade.

A typical trunk-amplifier spacing is 22 dB. Using 860-type coaxial cable for outside plant, we can go about 1700 ft. and use up the 22 dB at the top frequency of 750 MHz. Two amplifier spacings then would mean that we could locate the fiber-node location 3400 ft. away from the last building-entry location.

Not quite! We have to account for splitters, directional couplers for every building that we want to connect to the coaxial loop. Each splitter reduces the 1700 ft. spacing by about 300 ft. This means that from the fiber node we can go for instance

- 500 ft. to the 2nd building served
- 500 ft. to the 3rd building served
- we need the 2nd trunk station
        at the 3rd building served
- and we can repeat the 500 ft.
        or 500 ft. distances for the 4th and 5th building

What we have determined is that 5 buildings can be served as a group using coaxial cable provided that they meet the above distance limitations.

By locating the fiber node in the center of a group of buildings, we have to reduce the 2000 ft. distance to building No. 5 to 1700 ft., but we can now go in two directions and cover a group of 9 buildings.

Fig. I-5 shows an idealistic model of a coaxial network served by one fiber node. The first trunk amplifier is spaced approximately 1300 ft., or 433 yards, from the fiber-node location, which has been strategically placed in the center of the coaxial service area. As a result, an area of 600,000 square yards can be served using one 22-dB amplifier spacing. Using 1300 ft., again, to place the second trunk station, the coaxial service area can be extended to 2,350,000 square yards.

The real world, of course, is much different from this idealistic model.

Buildings are not located at prospective amplifier locations

Existing manholes and conduit sections dictate the routing of both fiber and coaxial cables

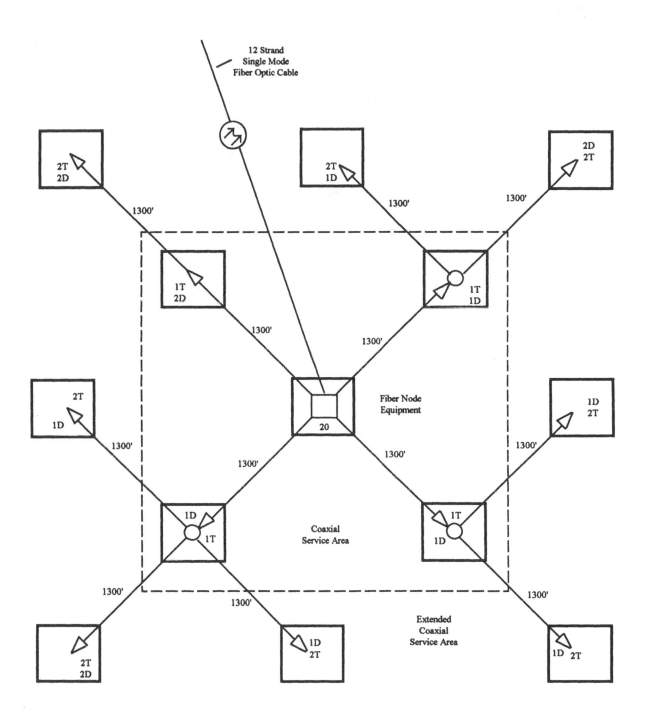

Fig. 1-5 Fiber Node Location Analysis

It is very much left to the ingenuity of the designer to decide the sizes of coaxial service areas. The conservative designer may use any one or all of the following ideas to determine the most cost-effective interface between fiber and coaxial service:

> Bring fiber to all building-entry locations
> (maximum cascade - 2 amplifiers)

> Bring fiber to buildings requiring two amplifiers in cascade
> and collect buildings, requiring only one amplifier at
> a fiber-node location, requiring one trunk-amplifier spacing
> (maximum cascade - 2 amplifiers)

> Utilize two trunk stations in cascade only for buildings with
> 1 amplifier (maximum cascade - 3 amplifiers)
> and utilize one trunk-station spacing for all 2-amplifier
> buildings
> (maximum cascade - 3 amplifiers)

> Bring fiber to only a few node points, use two trunk-station
> spacings for I and 2-amplifier buildings
> (maximum cascade - 4 amplifiers)

What are the differences between these different design approaches?

To quickly analyze the differences, it is appropriate to consult our idealistic campus model (Fig. I-5) one more time with respect to the number of buildings and the number of amplifiers within each building.

*   There are I3 buildings
*   Buildings fed from the first trunk station are marked IT
*   Buildings fed from the second trunk station are marked 2T
*   Single, internal distribution amplifiers are marked ID
*   2 Distribution amplifiers in cascade are marked 2D

Fiber to all Buildings

The model of Fig. I-5 can be used to assemble the following components:

*   I3 Fiber-optic cables (I2 strands each)
*   I3 Fiber-node equipment
*   O Feet of coaxial-cable construction

24

* O First trunk stations
* O Second trunk stations
* I3 First distribution amplifiers
* 5 Second distribution amplifiers
* 2 Maximum number of amplifiers in cascade
* I8 Total number of amplifiers
* 3 dB Carrier-to-noise degradation

The large number of I2-strand, single-mode fiber-optic cable and its installation is considered a substantial cost factor. Since we have not defined where and how far away the origination point of the fiber is located, no estimate can be made.

The electrical performance of the network is considered excellent with only the contribution of 2 amplifiers in cascade accounting for a 3-dB carrier-to-noise degradation. The total number of amplifiers is very low and, with a total of I3 amplifiers, does not represent a maintenance concern.

Conclusion: Highest cost
                 Excellent performance
                 Lowest maintenance

Fiber to Buildings with 2 Amplifiers
Coaxial Service to Buildings with I Amplifier

The model of Fig. I-5 can be used to assemble the following components:

* 8 Fiber-optic cables (I2-strand each)
* 8 Fiber-node equipment
* 9IOO ft. Coaxial cable
* 7 First trunk stations
* O Second trunk stations
* I3 First distribution amplifiers
* 5 Second distribution amplifiers
* 2 Maximum number of amplifiers in cascade
* 25 Total number of amplifiers
* 3 dB Carrier-to-noise degradation

The number of I2-strand, single-mode fiber-optic cables and its installation has been reduced to 8. The offset is 9IOO ft. of coaxial cable construction. Conduit placement and cable-pulling expenses are similar for fiber and coaxial cables.

Pulling a coaxial cable with the fiber cable reduces overall cable installation costs.

The electrical performance is considered excellent, considering that there are never more than 2 amplifiers in cascade.

The total number of amplifiers has increased from 13 to 20, but it still does not represent a concern for increased maintenance costs.

Conclusion: Second-highest cost
Excellent performance
Low maintenance

Fiber plus 1 Trunk-spacing to Buildings with 2 Amplifiers and 2 Trunk-spacings to Buildings with 1 Amplifier

The model of Fig. I-5 can be used to assemble the following components:

*   5 Fiber-optic cables (12-strand each)
*   5 Fiber-node equipment
*   11,700 ft. Coaxial cable
*   7 First trunk stations
*   3 Second trunk stations
*   13 First distribution amplifiers
*   5 Second distribution amplifiers
*   3 Maximum number of cascaded amplifiers
* 28 Total number of amplifiers
*   4-5 dB Carrier-to-noise degradation

This installation model would reduce the number of 12-strand fibers from 13 to 5. On the other hand, coaxial-cable construction is increased to 11,700 ft.

The electrical performance is still considered very good, even though the third amplifier in cascade has increased the noise buildup by 1 to 2 dB. A doubling of the number of cascaded amplifiers creates an additional 3 dB of noise. This means that 2, 4, 8, 16, 32 amplifiers in cascade would add 3, 6, 9, 12, 15 dB of noise to the video signal. It is obvious that 8, 16 or 32 amplifier cascades are not tolerable for carrier-to-noise ratios above 50 dB. Cascades of no more than 4 amplifiers, however, provide excellent picture quality with carrier-to-noise ratios over 50 dB.

For maintenance considerations, it is noted that the quantity of amplifiers has

26

increased to 23. Cable TV amplifiers have excellent availability records and the MTBF (Mean Time between Failure) is measured in years. Maintainability of this installation model is still considered high.

Conclusion: Third-highest, or second-lowest cost
                Very good performance
                Somewhat higher maintenance requirements

Fiber to selective Node Locations, 2 Trunk-Station Spacing and 2 Building Amplifiers for Coaxial Cable Service

The model of Fig. I-5 can be used as drawn to assemble the following components:

*   I Fiber-optic cables (I2 strands)
*   I Fiber-node equipment
*   I5,6OO ft. Coaxial cable
*   4 First trunk stations
*   8 Second trunk stations
*   I3 First distribution amplifiers
*   5 Second distribution amplifiers
*   4 Maximum numbers of amplifiers in cascade
*   3O Total number of amplifiers
*   6 dB Carrier-to-noise degredation

Only I single-mode, fiber-optic, I2-strand cable is required to serve a quite large area. Coaxial service is provided to I3 buildings using about I5,6OO ft. of cable. The maximum number of cascade has increased to 4, which will increase the transmission noise by another I or 2 dB. Very good carrier-to-noise ratio in excess of 5O dB, however, can be assumed. The number of amplifiers has increased to 3O, which only presents an increase of I2 over the quantity of I8 that was noted for fiber service to all buildings.

Conclusion: Lowest costs
                Very good performance
                Somewhat higher maintenance requirements

Network Design Groundrules

Based on the foregoing discussion of single buildings and campuswide multiple buildings, it appears that the designer has the freedom to apply his or her own preferences into the video network design. The previous discussion only offers a framework to the designer in which to apply personal ingenuity. The detailed design information is provided on a step-by-step basis in other chapters of this design guide.

What are the most important factors to be considered in the planning process?

a) Broadband coaxial cable is the prefered medium for high-capacity, bi-directional transmission networks in single buildings of any complexity

b) Campuswide video networks should incorporate single-mode fiber cables to strategically selected fiber-node locations. The translation to coaxial cable provides for a cost-effective installation

c) The buildings served by coaxial cable shall not have more than 4 amplifiers in cascade, which includes a maximum of 2 amplifiers in cascade within a building

d) Single-mode fiber cable for video network distribution shall only be considered if and when the number of cascaded amplifiers in a coaxial system would exceed three

e) Bringing fiber to every building-entry point is a more expensive proposition and may not offer better network performance. Careful planning and the establishment of comparison routing plans must be undertaken before designing the network. These comparison routing plans must compare equipment, construction and installation cost factors to determine the most cost-effective and practical solution

f) The designer must use common sense as the predominant guide to accomplish the design task. Vendor and supplier publicities are produced to create confusion and promote security in an unknown future. The designer is encouraged to see through the fluff and determine concrete steps, on a solid foundation of common sense, towards a realistic solution for the future.

Chapter 2

Spectrum Utilization

The Capacity of a Broadband Coaxial Cable

A single coaxial cable has an enormous channel capacity. When we are talking about a channel, we are referring to a 6-MHz video channel capable of carrying video, stereo-audio and chroma carrier.

The transmission of an NTSC video channel, usually and until now, consists of an analog signal stream that can be modulated to any frequency assignment on the cable. Cable modulators are fairly inexpensive and can be used to stack one channel on top of another. The TV-set at the receiving location utilizes an inexpensive demodulator or tuner to remove the carrier frequency and to separate video and audio for display on the screen. Cable-ready TV-sets incorporate tuners capable of selecting 100 or more channel assignments.

One broadband coaxial cable can transmit these 100 channels at the same time. This capability is also often referred to as the "spectrum".

The design of your corporate broadband video network must include your own rationale relative to the "utilization" of this vast spectrum.

If a data network transmits 100 Mbit/s data rates, we are quite impressed with the speed of such a transmission and that a number of data streams can be transmitted simultaneously. When we analyze the 100 Mbit/s speed for video transfer, we find that in its uncompressed state a single full-motion video transmission requires 100 Mbit/s for real-time transfer.

Based on this high-speed requirement for video, it appears unlikely that multichannel,
bi-directional video transmission can be established on a data network.

The broadband coaxial cable, on the other hand, has the capacity to transmit over 100 video channels simultaneously and within a frequency spectrum of 5 to 750 MHz or even to 1000 MHz.

This capacity can be obtained using standard and inexpensive cable components to feed many users in a tree-and-branch architecture that only requires one backbone cable. FDDI, or even ATM technologies, would require massive twisted-pair or fiber bundles to achieve the same handling capacity as well as an ATM switch of enormous proportions.

The spectrum of the broadband coaxial system can be arranged for forward and return transmission segments. The more common coaxial broadband architectures are discussed in the following:

a)   The Sub-Split System
b)   The Mid-Split System
c)   The High-Split System
d)   The Dual-Cable System
e)   Other Evolutionary Configurations

The Sub-Split System

The Sub-Split System is the conventional distribution system architecture of the cable industry.

The emphasis of a cable-TV delivery system is the forward direction. Return-transmission capability exists in a limited manner and has not been utilized, except in isolated cases.

The forward capacity of the Sub-Split System covers the frequency band from

54 to 550 MHz, or even 750 MHz. Since a video channel uses a 6-MHz space, a total of 78 or III simultaneous video transmissions can be conducted.

The return capacity of the Sub-Split System, however, is limited to the 5 to 30 MHz passband, which only permits the simultaneous transmission of 4 video channels.

The sub-low band from 5 to 30 MHz is also subject to a variety of interferences ranging from shortwave transmission, citizen's band to medical equipment spurious transmission products. For these reasons, the sub-low return band is sometimes used for data transmissions and status monitoring, but very rarely for video.

It is concluded that the Sub-Split System is the preferred architecture for the delivery of a large number of video channels in the forward direction, but that it is not usable for multiple-channel return transmissions. The Sub-Split System then becomes the ideal architecture for a residential or entertainment TV delivery system, but does not support many two-way interactive transmission requirements of an enterprise network.

Sub-Split Spectrum Utilization Considerations

The 5-30 MHz Spectrum

The 5-30 MHz spectrum can be used for return transmissions. The band consists of 4 6-MHz frequency assignments. These 24 MHz can also be used for high-speed data and telephony assignments. For instance, six T-I transmissions can be arranged within one video channel of 6 MHz.

The 30-54 MHz Spectrum

This frequency band is utilized by the crossover filter that separates forward and return transmissions and cannot be used for video transmission.

The 54-88 MHz Spectrum

This band contains the original broadcast TV-transmission frequencies (Ch 2 through Ch 6). These five channels are conventionally used for entertainment television.

The 88-108 MHz Spectrum

The 88-108 MHz band is assigned by the Federal Communications Commission (FCC) for FM broadcast transmission. This band should be reserved for FM radio on cable or data transmission. TV transmission in this frequency band would be interfered with by strong neighboring FM broadcast stations.

The 108-120 MHz Spectrum

Due to Federal Aviation Administration (FAA) use of these frequencies, it is recommended by the FCC that this band requires special egress limitations or not be used at all.

The 120-750 MHz Spectrum

This frequency band consists of 105 channels that can be assigned for forward video transmission. Considering the 6 channels in the low band, the total number of forward channels is 111.

Two-way Transmission

It is recommended that forward data communications be located in the 246-276 MHz spectrum. This 30 MHz of spectrum follows IEEE 802.4 guidelines for broadband LANs and can be used in conjunction with the 5-30 MHz return-transmission band.

Conclusions

The Sub-Split Spectrum is ideally suited for forward entertainment television requiring channel assignments starting at Ch 2 - Ch 6 and continuing with Ch 7 - Ch 112.

The assignment of spectrum segments for services such as

> Local channels
> Satellite channels
> Corporate Information channels
> Video Retrieval channels
> Video-on-Demand channels
> Data or Voice T-1 channels

are left to the desires of the system designer. It is recommended that channel allocations are determined at the beginning of the system-design process. An example of an ETV Entertainment system channel spectrum assignment plan is provided in Fig. 2-1, in the right column. The spectrum assignment stops at Ch 77 at 540 MHz, but can be extended to Ch 111 at 750 MHz, or even to Ch 151 at 996 MHz.

The Mid-Split System

Mid-Split Systems have been provided by some cable companies to fulfill the franchise requirements for an institutional interconnect network.

A Mid-Split System has a forward spectrum of 160-550 or 750 MHz, which provides for a channel capacity of 65 or 98 channels.

The return capacity of the Mid-Split System utilizes the frequency band from 5-112 MHz. A casual look at this spectrum indicates that 17 simultaneous return-channels are available. However, when eliminating the lower part of the spectrum (5-30 MHz) for interference reasons and the 88-108 MHz band, which is used by local FM broadcast stations as a potential interference source, the number of usable return-channels shrinks to about 8 channels.

It is concluded that the Mid-Split System can be used in a limited manner for residential and instructional services. By assigning two-way enterprise services to the lower portion of the spectrum, the IEEE 802.4 transmission assignments can be met and residential services can be provided on Ch 45 or higher.

The negative aspect of providing residential and academic services on one cable lies in the fact that students cannot watch television starting at Ch 2 and that all TV-sets must be tunable and cable-ready for reception of higher number channels.

Mid-Split Spectrum Utilization Consideration

The 5-112 MHz Spectrum

The return or inbound transmission can utilize the 5-112 MHz band. Division by six seems to indicate that there are 17 inbound channel spaces available.

A closer analysis of the 5-112 MHz spectrum, however, shows some concerns relative to the quality of transmission.

The sub-low band from 5-50 MHz is subject to a variety of interferences ranging from short-wave transmission, citizen's band, ham transmission to medical equipment interference. The use of sub-low frequencies has been usually restricted to data and status-monitoring transmissions, but could be used for security video channels.

The FM band is located in the 88-108 MHz band. Any local FM station will

Channel Designation (Historical)	Channel Designation (EIA)	BAND	Frequencies — Standard	Frequencies — IEEE 802.4
T7	N/A	SUB LOW OR T BAND	5.75	5.75
T8			11.75	11.75
T9			17.75	17.75
T10			23.75	23.75
T11			29.75	29.75
T12			35.75	35.75
T13			41.75	41.75
T14			47.75	47.75
2	2	LOW BAND	54	2' 53.75A
3	3		60	3' 59.75B
4	4		66	4' 65.75C
4+, A-8	1		72(2)	4A' 71.75D
5, A-7	5		76(2)	5' 77.75E
6, A-6	6		82(2)	6' 83.75F
FM1, A-5	95	FM BAND	90	FM1' 89.75
FM2, A-4	96		96	FM2' 95.75
FM3, A-3	97		102	FM3' 101.75
A-2	98	MID BAND	108	
A-1	99		114	
A	14		120	
B	15		126	
C	16		132	
D	17		138	
E	18		144	N/A
F	19		150	
G	20		156	
H	21		162	
I	22		168	
7	7		174	
8	8		180	
9	9	HIGH BAND	186	
10	10		192	
11	11		198	N/A
12	12		204	
13	13		210	
J	23	SUPER BAND	216	216
K	24		222	222
L	25		228	228
M	26		234	234
N	27		240	240
O	28		246	246A
P	29		252	252B
Q	30		258	258C
R	31		264	264D
S	32		270	270E
T	33		276	276F
U	34		282	282
V	35		288	288
W	36		294	294
AA	37	HYPER BAND	300	
BB	38		306	
CC	39		312	
DD	40		318	
EE	41		324	
FF	42		330	
GG	43		336	
HH	44		342	
II	45		348	
JJ	46		354	
KK	47		360	
LL	48		366	
MM	49		372	N/A
NN	50		378	
OO	51		384	
PP	52		390	
QQ	53		396	
RR	54		402	
SS	55		408	
TT	56		414	
UU	57		420	
VV	58		426	
WW	59		432	
XX	60		438	
YY	61		444	
ZZ	62		450	
	63		456	
	64		462	
	65		468	
	66		474	
	67		480	
	68		486	
	69		492	
N/A	70		498	N/A
	71		504	
	72		510	
	73		516	
	74		522	
	75		528	
	76		534	
	77		540	

ITV Instructional or Academic System

- Unused
- Reserved for High Speed Data, Energy Management and Security Video (10) Channels
- Unused
- Simultaneous Remote Video and Teleconference Transmissions (13) Channels
- High Split Crossover Area
- Reserved for Energy Management, Tele-conferencing and High Speed Data Circuits (11) Channels
- Generated (5) channels
- Channel Expansion (4) channels
- Off-Air Satellite (5) channels
- Expansion Off-Air-Satellite (4) channels
- VRS Transmission channels with Classroom Control
- Video Retrieval System (16) Channels Reserved
- Reserved for On-Demand TV Services

ETV Entertainment System

- Optional Data Return (4) channels
- Sublow Crossover Area
- Entertainment Service (5) channels
- Optional FM Stereo
- Unused
- Entertainment Service — Total Number (22) Satellite and (12) Off-Air Channel Initially (29) channels Total (34) channels CH2. through CH.34
- Generated (5) channels
- Channel Expansion (4) channels
- Entertainment Service Expansion Area (20) channels
- Reserved for Student On Demand TV Services (13) channels

Fig. 2-1 Spectrum Utilization Example - Sub-split and High-split Networks

34

introduce beat products in a video transmission that are difficult to combat even with the tightest radiation avoidance controls.

The remaining frequency band of 50-88 MHz is commonly referred to as the Low Band. It consists of Ch 2 to Ch 6 transmission.

Local high-powered broadcast stations can ingress into the cable and provide ghosting and/or co-channel images.

As a result of this analysis, it appears that, at best, there are only five (5) video channels available simultaneously for high-quality, interference-free, inbound transmissions, one channel reserved for high-speed data and (8) channels for occasional data and security video, for a total of (13) channels.

The 112-150 MHz Spectrum

This is the crossover band of the Mid-Split System. Frequencies below 112 MHz are inbound. Frequencies above 150 MHz are outbound. There are no problems with FAA frequencies in this system because there are no frequencies transmitted or received in this band.

The 150-750 MHz Spectrum

This frequency band consists of 100 channels that can be assigned for forward video transmission. This spectrum can be subdivided in accordance with the desires of the designer or the communications manager.

Fig. 2-2 shows a mid-split spectrum arrangement that separates academic and residential services from each other. Resident or auxiliary entertainment services are only provided over 306 MHz. Academic and bi-directional services to classrooms and presentation rooms are restricted to the 150-336 MHz band. Filters are provided to separate the services. Students in dormitories can only receive entertainment and video-on-demand channels. Academic locations cannot receive entertainment channels unless they are selected for educational use.

Two-way Transmission

Again, it is recommended that forward, high-speed data communications be located in the 246-276 MHz spectrum. This 30 MHz wide spectrum follows IEEE 802.4 guidelines for broadband LANs and can be used in conjunction with offset low-band channel assignments. Data traffic sent towards the headend or gateway at anyone of the low-band frequency is translated at the headend applying a 192.25 MHz translation frequency. This translation frequency changes the incoming signals into forward transmission frequencies between 246 and 276 MHz.

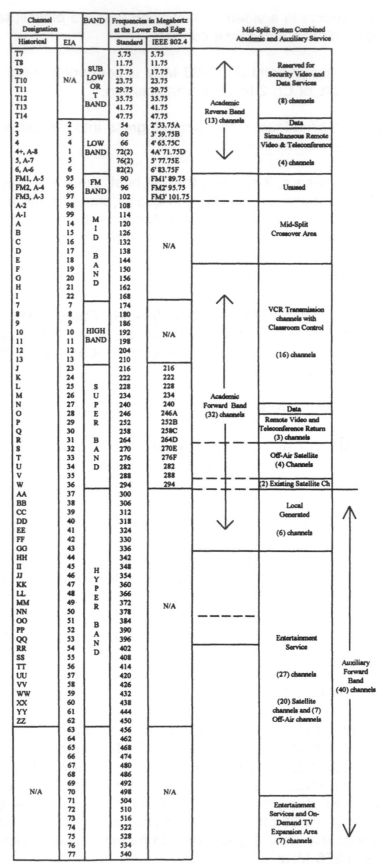

Channel Designation		BAND	Frequencies in Megahertz at the Lower Band Edge		Mid-Split System Combined Academic and Auxiliary Service
Historical	EIA		Standard	IEEE 802.4	
T7			5.75	5.75	
T8			11.75	11.75	Reserved for Security Video and Data Services
T9		SUB LOW OR T BAND	17.75	17.75	
T10	N/A		23.75	23.75	(8) channels
T11			29.75	29.75	
T12			35.75	35.75	
T13			41.75	41.75	
T14			47.75	47.75	
2	2		54	2' 53.75A	Data
3	3		60	3' 59.75B	Simultaneous Remote Video & Teleconference
4	4	LOW BAND	66	4' 65.75C	
4+, A-8	1		72(2)	4A' 71.75D	
5, A-7	5		76(2)	5' 77.75E	(4) channels
6, A-6	6		82(2)	6' 83.75F	
FM1, A-5	95	FM BAND	90	FM1' 89.75	
FM2, A-4	96		96	FM2' 95.75	Unused
FM3, A-3	97		102	FM3' 101.75	
A-2	98		108		
A-1	99		114		
A	14	MID BAND	120		Mid-Split Crossover Area
B	15		126		
C	16		132	N/A	
D	17		138		
E	18		144		
F	19		150		
G	20		156		
H	21		162		
I	22		168		
7	7		174		VCR Transmission channels with Classroom Control
8	8		180		
9	9		186		
10	10	HIGH BAND	192	N/A	
11	11		198		
12	12		204		(16) channels
13	13		210		
J	23		216	216	
K	24		222	222	
L	25	SUPER BAND	228	228	
M	26		234	234	
N	27		240	240	Data
O	28		246	246A	Remote Video and Teleconference Return (3) channels
P	29		252	252B	
Q	30		258	258C	
R	31		264	264D	
S	32		270	270E	Off-Air Satellite (4) Channels
T	33		276	276F	
U	34		282	282	
V	35		288	288	
W	36		294	294	(2) Existing Satellite Ch
AA	37		300		
BB	38		306		Local Generated
CC	39		312		
DD	40		318		
EE	41		324		(6) channels
FF	42		330		
GG	43		336		
HH	44		342		
II	45		348		
JJ	46	HYPER BAND	354		
KK	47		360		
LL	48		366		
MM	49		372	N/A	
NN	50		378		
OO	51		384		
PP	52		390		
QQ	53		396		Entertainment Service
RR	54		402		
SS	55		408		
TT	56		414		(27) channels
UU	57		420		
VV	58		426		
WW	59		432		
XX	60		438		(20) Satellite channels and (7) Off-Air channels
YY	61		444		
ZZ	62		450		
	63		456		
	64		462		
	65		468		
	66		474		
	67		480		
	68		486		
	69		492		
N/A	70		498	N/A	Entertainment Services and On-Demand TV Expansion Area (7) channels
	71		504		
	72		510		
	73		516		
	74		522		
	75		528		
	76		534		
	77		540		

Fig. 2-2 Spectrum Utilization Example - Mid-Split Network

The same frequencies, or any other 6 MHz channel pairs, can be used for interactive television, distance learning or desktop videoconferencing assignments.

Conclusions

The Mid-Split spectrum can serve a large number of one-way entertainment and academic channels as well as a limited number of two-way channels. It can be anticipated by the designer that if there will be the need for more than I3 bi-directional video transmissions at any time in the future, a move towards the High-Split system may be considered.

The assignment of the spectrum segments for the various services such as

    Local channels
    Satellite channels
    Corporate Information channels
    Video Retrieval channels
    Video-on-Demand channels
    Desktop Videoconferencing channels
    Data or Voice T-I channels

are left to the desires of the system designer, the communications manager and the user population.

If the Mid-Split facility is used for both academic and residential services, it is important to locate the academic band in the lower portion of the forward spectrum. The reasons for this recommendation are two-fold. First, all data assignments are located in the 246-276 MHz frequency band. Second, it is desirable to locate residential services in a continuous manner on the dial. When channel assignments are used over 276 MHz, all channel numbers are consecutive from Ch 34 to Ch 99 on a cable-ready TV set.

The High-Split System

Very similar to the Mid-Split system in design, the High-Split system offers a crossover region at higher frequencies. Forward transmission is limited to frequencies above 222 MHz. Return transmissions are possible in the frequency spectrum of 5-I86 MHz, for a total of 3O channels. Even, when considering that IO channels may be unusable because of interference, a total of 2O channels are available for high-quality video transmissions.

This high-channel capacity in the return direction makes the High-Split system an ideal choice for a flexible and obsolescence-free two-way video system.

As in the case of the Mid-Split system, the High-Split system can be used for combined residential and academic services. However, the same negative aspects relative to the tuning of student TV-sets exists. In addition, both Mid-Split and High-Split systems deliver academic and residential services to all outlets unless expensive filtering is provided to separate these services.

With a return-capacity of 30 simultaneous 6 MHz channels, the High-Split architecture may be the choice of both cable and telephone companies in future HFC distribution networks. While the sub-low band only accommodates low and high-speed digital traffic from residential homes, the High-Split architecture would permit analog NTSC and digital video transmission from multimedia workstations and computer LANs.

In the environment of an enterprise, the High-Split coaxial network will provide ample capacity for interactive TV in analog and in uncompressed form for the foreseeable future.

High-Split Spectrum Utilization Considerations

The 5-186 MHz Spectrum

The return or inbound transmissions can be distributed in this band. Division by six seems to indicate that there are 30 inbound video channels available.

Considering the same cautions that were expressed in the discussion of the Mid-Split system, the capacity for high-quality video transmissions is at least increased from 5 to 17. This represents a three-fold increase of simultaneous high-quality video return-channels.

The 186-222 MHz Spectrum

This is the crossover-band of the High-Split system architecture. Frequencies below 186 MHz are inbound. Frequencies above 222 MHz are outbound. Return transmissions in the 108-120 MHz band have to be carefully selected to avoid interference with FAA air-to-ground frequencies. A tight, well designed and constructed coaxial broadband network is required to minimize the dangers of egress.

The 222-750 MHz Spectrum

The forward spectrum of the High-Split cable consists of 88 video channels or 6 MHz frequency assignments. With 30 channels reserved for two-way or bi-directional voice, data or video traffic, there are an additional 58 channels available for the forward distribution of television. Again, the system designer is at liberty to subdivide the spectrum in accordance with existing and future service requirements.

Fig. 2-I indicates an example of the spectrum utilization of a High-Split system in the center column. The column is marked "ITV Instructional or Academic System". The chart only carries the frequency assignment to Ch 77 or 540 MHz. A 750 MHz network design would add 35 additional channel assignments.

Again, as in the case of the Mid-Split system of Fig. 2-2, the forward-transmission band could be separated between academic two-way and residential one-way services. For instance, frequencies from 222-450 MHz could be assigned to satisfy 30 bi-directional and 8 forward channels for classroom and/or academic locations. The remaining 300 MHz band or 50 channels could be used for entertainment and educational channels to dormitories, residencies etc.

Fig. 2-3 indicates a check-list for voice, data and video services that can be used by the designer to conduct frequency spectrum assignment studies and to develop the most appropriate selection of frequencies for immediate and future services.

The Dual-Cable System

The optimum solution for a campuswide video network, the one which offers spare capacity to satisfy any future requirements for additional spectrum, is the Dual-Cable system.

Fig. 2-I shows the spectrum capacity of both the residential Sub-Split system (ETV) and the academic High-Split system (ITV).

The ETV, or residential system, has a passband of 54-550 MHz for 77 forward video channels. If designed for an upper frequency of 750 MHz, the ETV residential system can provide for up to 111 simultaneous video channels.

The ITV, or academic system, is a High-Split system with a return frequency range of 5-186 MHz, or 30 channels, and a forward frequency range of 222-550 or 750 MHz for 54, or even 88, simultaneous video channels.

Fig. 2-I shows the spectrum architecture of a Dual-Cable system relative to channel-assignment designations of EIA, Standard and IEEE 802.4. The residential or ETV system features continuous forward channel assignments between 54-550 MHz. A total of 77 channels can be delivered to all residential outlets starting with Ch 2. Depending upon future expansion requirements, system design can also take the passband up to 750 MHz, providing for an additional 33 channels.

Channel Designation Historical	EIA	BAND	Frequencies in Megahertz at the Lower Band Edge Standard	IEEE 802.4
T7			5.75	5.75
T8			11.75	11.75
T9		SUB LOW OR T BAND	17.75	17.75
T10	N/A		23.75	23.75
T11			29.75	29.75
T12			35.75	35.75
T13			41.75	41.75
T14			47.75	47.75
2	2		54	2' 53.75A
3	3		60	3' 59.75B
4	4	LOW BAND	66	4' 65.75C
4+, A-8	1		72(2)	4A' 71.75D
5, A-7	5		76(2)	5' 77.75E
6, A-6	6		82(2)	6' 83.75F
FM1, A-5	95	FM BAND	90	FM1' 89.75
FM2, A-4	96		96	FM2' 95.75
FM3, A-3	97		102	FM3' 101.75
A-2	98		108	
A-1	99		114	
A	14	MID BAND	120	
B	15		126	
C	16		132	
D	17		138	
E	18		144	N/A
F	19		150	
G	20		156	
H	21		162	
I	22		168	
7	7		174	
8	8		180	
9	9		186	
10	10	HIGH BAND	192	
11	11		198	N/A
12	12		204	
13	13		210	
J	23		216	216
K	24		222	222
L	25	SUPER BAND	228	228
M	26		234	234
N	27		240	240
O	28		246	246A
P	29		252	252B
Q	30		258	258C
R	31		264	264D
S	32		270	270E
T	33		276	276F
U	34		282	282
V	35		288	288
W	36		294	294
AA	37		300	
BB	38		306	
CC	39		312	
DD	40		318	
EE	41		324	
FF	42		330	
GG	43		336	
HH	44		342	
II	45		348	
JJ	46		354	
KK	47		360	
LL	48	HYPER BAND	366	
MM	49		372	N/A
NN	50		378	
OO	51		384	
PP	52		390	
QQ	53		396	
RR	54		402	
SS	55		408	
TT	56		414	
UU	57		420	
VV	58		426	
WW	59		432	
XX	60		438	
YY	61		444	
ZZ	62		450	
	63		456	
	64		462	
	65		468	
	66		474	
	67		480	
	68		486	
	69		492	
N/A	70		498	N/A
	71		504	
	72		510	
	73		516	
	74		522	
	75		528	
	76		534	
	77		540	

Spectrum Assignment — Voice | Data | Video

Status Monitor

Data — 802.4 Preferred

Voice or Data

(11) In-bound 2-way Video

High-Split Cross Over Area

Status Monitor

Data — 802.4 Preferred

Voice or Data

(11) Outbound 2-way Video

One-Way Outbound Video (32 or 66 channels)

Fig. 2-3 Frequency Spectrum Assignment- High-split - Check List

The capacity of the residential system permits the development of a comprehensive channel plan for entertainment and education channels as well as enterprise-originated channels and student video-on-demand programs.

Through switching equipment at the headend, all or some enterprise-originated channels can appear on both the residential and the academic systems.

It is emphasized that the Dual-Cable system requires a campus environment to be a cost-effective solution. On a campus of a University, for instance, buildings are grouped for housing and educational services. The architecture of the Dual-Cable system would separate the network and serve residential buildings with a sub-low cable and the academic buildings with a high-split cable. In this architecture are only a few areas where dual-coaxial cables are required and the cost increase over a single system is minimal.

If, on the other hand, academic and residential services are mixed within the same building, the cost of establishment of a Dual-Cable system may be substantial because of dual-cable requirements to some floors, dual risers and duplication of cable and equipment.

Other Spectrum Utilization Approaches

Many developers of HFC systems worry about the utilization of the available I GHz spectrum. Since there are no standards for spectrum utilization in the outside world, the development of cable modems has been rather slow.

In Time Warner's Orlando, Fla., trial, for example, the cable operator allotted bandwidth for two-way services from 650 MHz to I GHz, with 250 MHz available for duplex communications. Of that, about I00 MHz of the system goes to wireless personal communications services.

The frequency plan proposed by Scientific Atlanta uses also a I GHz coaxial telecommunications spectrum.

The I GHz network includes: 400 MHz for analog video, 250 MHz for digital data and 300 MHz for communications.

According to some vendors, the demand in the U.S. for a reverse channel bandwidth limitation to about 5-42 MHz, depends greatly on how interactive the system is meant to be.

These spectrum utilization schemes are all based on the outside world and the growth of residential traffic. It is foreseen that future residential traffic consists of voice and data from the PC and the set-top converter, and that downstream transmission consists of multiple video-on-demand channels. This downstream orientation may be obsolete in a few years when PCs are connected to broadband networks with CD-ROM and multimedia boards.

But we should not worry too much about the development in the outside world or of the information superhighway. Within the enterprise, multiple bi-directional video transmission can be provided now and meet any future growth requirements, using established broadband coaxial cable spectrum architectures.

Chapter 3

Analog and Digital Video Transmission

The Video Universe

Broadcast Television

The beginning of television in the U.S. was based on the RS I7O standard that covers the transmission of video and audio in analog form and within a 6 MHz frequency band.

Commonly referred to as NTSC video, the analog standard sets the frequency of the video carrier I.5 MHz inside the 6 MHz band with the audio carrier at 4.5 MHz.

Despite the single-sideband transmission of all video information within about a 4.2 MHz band, the broadcast quality of the video has been generally satisfactory to most consumers. Because of interference between frequency allocations of adjacent channels, the FCC licensed nonadjacent channels first in the low and high VHF band, Ch 2 through Ch 13, and later in the UHF band.

The nonadjacent channel allocation stems from the inability of the TV-set tuner to discern two adjacent broadcast channels. For this reason, the FCC allocated channel numbers 2,4,5 and 7,9,11, 13 first in the largest metropolitan areas. Intermediate metropolitan areas were assigned interstitial channels such as Ch 3, 6,8,10 and 12. To accommodate additional broadcast licensees, the FCC opened the UHF band with an array of channel allocations between Ch 19 and 83. Even though, there are many UHF channels in service at the present time, the band has never been fully subscribed. The reason for this underutilization of the UHF band is directly related to the reduced coverage area of a broadcast transmitter at the higher frequencies of the UHF band.

NTSC composite analog video was initially developed with separate video and sync signals. Video was distributed without sync as a 1 V peak-to-peak (Vpp) signal. Black was 0 V and peak white was 1 V. Sync signals were distributed at 4 Vpp. Both signals were usually clamped to establish a reference black level at 0V, peak white at 1 V and sync tip at -4 V. When video and sync signals were combined into a single composite signal for transmission, sync amplitude was reduced by a factor of 10. The resulting signal was 1.4Vpp with the sync tip at -0.4 V. At this point, the precedent of a 10:4 video-to-sync ratio was established. Later, when the overall signal amplitude was reduced to 1 Vpp, the 10:4 ratio remained.

The IRE (Institute of Radio Engineers, later the IEEE) established a unit of measure for video signals that was 1/140 of Vpp or 714 mV. 100 IRE units represent the video information and has a 714 mV amplitude. The sync is represented by -286 mV. The sum of the two is the conventional video level of 1 Vpp. The video information then is amplitude modulated to the carrier frequency that is licensed to the broadcaster. Fig. 3-1 shows the relationship of the IRE scale and the timing of a video frame.

Cable Television

The development of the three networks and the growing number of broadcast TV stations led to the formation of Community Antenna Television systems (CATV), first in rural and over-the-horizon communities and later, with the advent of satellite technology, in suburbia and urban areas.

Early cable TV systems, in the late 1950s, carried a maximum of five channels. Systems in the 1970s carried 30 channels in the frequency band of 54 to 300 MHz.

Present-day technology permits the transmission of about 150 channels in the

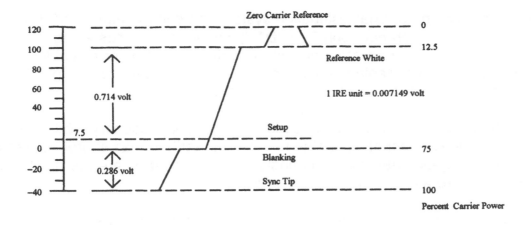

The IRE scale versus 1V peak to peak

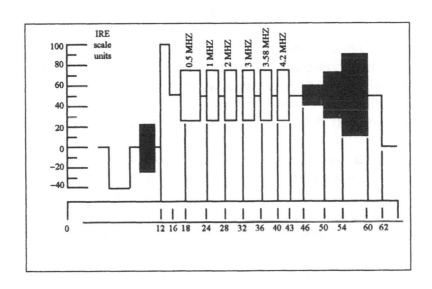

Fig. 3-1 NTSC Video Signal Standard and multiburst test signal
for frequency response test and chrominance non-linear phase and gain

frequency band of 54 to 1000 MHz. Each video and audio transmission is transferred to a cable channel assignment using a modulator. The precise application of multiplexing permits the "stacking" of many 6 MHz channels on the coaxial cable distribution network without any problems.

The television set or the set-top converter permits the selection of any one of the channels for display of a broadcast TV picture.

Due to the proliferation of cable TV during the past decade, cable and equipment costs are low. This cost-effectiveness of the RF broadband coaxial distribution system has led to about 60% of households being connected to cable, with over 90% of all households featuring at least one TV set.

The RF broadband coaxial distribution system can carry compressed digital standard TV and high-definition TV (HDTV) whenever it comes along. It also permits the transmission of voice and data in analog or digital form, which makes coaxial cable the prime choice for hybrid fiber coaxial (HFC) voice/data/video transmission systems of the future.

Since the inception of the satellite-based repeaters in stationary orbits, the number of video channels available to cable subscribers has sharply increased. In addition, pay-per-view event channels are multiplying as are the number of movies that are available for cable distribution.

To order a particular movie at 7:30 pm and not being able to select this movie again until the next day, or next Friday, at 11:00 pm, reduces the growth and, therefore, the added monthly revenues. The cable industry is limited in channel capacity. Video-on-demand (VOD) services can only be established when the subscriber is in the position to select a particular program at any time and to control the movie as if it is playing on a VCR.

The need for additional channels forces both the cable companies and telephone operating companies to reassess the distribution plant architecture. Existing cable television distribution systems are using many coaxial trunk and distribution amplifiers in series. As a result, the quality of transmission is impaired by noise and distortion components of cascaded amplifiers.

HFC offers an ideal solution to remedy both problems. From the headend, the fiber-optic cable is extended to a fiber-node location that covers up to 500 subscribers with coaxial distribution. Using this approach, the number of amplifications has been reduced to a range of 2 to 5, thus drastically reducing the effect of noise and distortion. In addition, the entire channel capacity of the system is now available to each group of up to 500 subscribers. This means that HFC will provide VOD capabilities much more readily than any other architecture.

Wireless TV Distribution Systems

Multichannel, Multipoint Distribution Service (MMDS)

In competition with coaxial cable delivered multichannel video systems, wireless TV distribution systems are emerging in many cities and surrounding areas.

Wireless TV distribution systems are based on line-of-sight microwave transmission and can deliver up to 60 channels of television directly to residential buildings. Even though, monthly subscriber rates are slightly lower, a receive location requires a special receiving antenna and decoding equipment that increases the complexities of the installation. The value of wireless transmission is directly related to the geographic coverage area, in as much as it can serve rural areas not presently wired by cable TV.

The negative aspect of wireless TV distribution systems may be the high cost of connecting a second television set and the inability to accommodate return-transmissions from the subscriber. This means that two-way services such as data and voice cannot easily be implemented.

In addition, it is noted that video-on-demand (VOD) programming cannot live with the limited spectrum of a wireless cable distribution system.

Wireless cable TV service is also called MMDS. MMDS is an abbreviation for multichannel multipoint distribution service. The multichannel designation was tacked onto the front of MDS, the system's single-channel predecessor, when it expanded to 31 channels.

In a wireless system, as many as 33 television channels are grouped together and broadcast simultaneously as a single block. These channels are broadcast in the 2.5-2.7 GHz band or in a block of 49 channels in the 27.5-29.5 GHz band. These bands of carrier frequencies are also referred to as the radio frequency.

Note that these carrier frequencies are ten and one-hundred times higher than those of standard VHF frequencies. This permits the use of very efficient high-gain, yet compact antennas. The receiver antenna sizes are less than 500 and less than 150 millimeters in diameter.

A low-noise block or down-converter moves or translates that block of channels down to the VHF or UHF bands. On the other hand, a set-top converter descrambles the signal and allows subscribers to select the desired channel. Due to the limited number of channels, it is expected that MMDS expansion is restricted to areas not wired by the cable companies.

Cellular Television

To combat this disadvantage, a new cellular TV distribution system has been established by the FCC in the 28 GHz band. The channel capacity will increase drastically by using these millimetric waves. The 27.5 -29.5 GHz Ka-band allows a 2000 MHz band-width, which, even when split into two sub-spectrums, can provide 150 channels of analog 6 MHz video.

The drawback of millimetric wireless transmission of video is the small coverage area. Cellular telephony presently uses transmit towers every 8-10 miles and relies on coverage areas of 5 to 6 miles. Cellular television at 28 GHz would reduce this coverage area to 1 or 2 miles and require a much denser installation of transmitter support structures.

Cellular TV can be delivered in analog form to satisfy the existing population of television sets or could be a digital transmission conforming to new MPEG-2 (Motion Picture Experts Group) global transmission standards.

Direct Broadcast Satellite Service (DBS)

Heralded since the inception of high-power satellite technology, DBS satellite-delivered multichannel TV distribution has been talked about since 1985.

Direct TV launched an HS-601 satellite into orbit on December 17, 1993 and will utilize 11 of the 16 transponders, for a total of 50 channels. The remaining transponders are used by Hubbard Broadcasting's USSB (United States Satellite Broadcasting). A second satellite launched in May of 1994, which will add 100 channels to Direct TV's programming, for a total capacity of 150 channels.

For the first time, digital television transmission using compression technology is being provided. The receiver has ten-times the power of a personal computer and delivers CD-quality audio and laserdisk-quality video. It processes complex authorization and security control data, stores pay-per-view selection habits, converts the TV signal from digital to analog and provides a 16:9 aspect ratio picture on wide-screen TV sets. The antenna is an 18-inch parabolic dish.

The digital DBS television delivery system is upgradable to HDTV (High Definition Television), will provide all presently known cable TV channels, pay-per-view movies, special sports events and special interest channels.

With the entire geographical area of the U.S. covered, DBS may become a formidable competition to cable TV, especially when the quality of the digital TV outperforms the typical cable TV reception.

The negative aspects of DBS are, again, the high costs of the antenna and of connecting a second independent TV set. Also, the inability to accommodate return transmission of two-way services such as data and voice is a negative factor.

The initial service opened using MPEG-1 (Motion Picture Experts Group) compression technology. MPEG-1 reduces the video bandwidth to 150 Kbit/s, which is not sufficient to represent full-motion video. DBS plans to change to MPEG-2 during the immediate future to combat the complaints of subscribers watching sports and other fast-moving background images.

Closed Circuit Television (CCTV)

Mostly concentrated within enterprises, closed circuit TV, or CCTV, describes one-way and two-way video transmission commonly utilizing baseband frequencies.This means that all CCTV systems are single-channel systems and only one picture can be viewed at a time.

The most common examples of CCTV systems are video production facilities and industrial security systems. The former represent the high-end, the latter the low-end of the range of CCTV. Security systems feature a multitude of camera locations, which are automatically switched for viewing by time sequence. Alarms are based on changes in the viewing field.

CCTV systems have been using analog black and white technology since the inception of video surveillance. The recent increased requirements for city and countywide traffic flow control systems have changed all that. Here, cameras are deployed to identify the number of cars in each of the four directions of an interchange. The analog image is immediately digitized and the digital image of the cars transmitted via telephone system to the evaluating computer center. Depending on the traffic density, the green/red sequence is changed by the computer to ease traffic congestions.

Traffic monitoring and control systems do not require the transmission of video as a composite picture, but illustrate the importance of the video image as the collector for digital transmission over standard telephone lines.

Instructional Two-Way Television

Due to the large channel capacity of coaxial cable systems, many bi-directional or two-way systems have been built by cable companies to interconnect public institutions.

The so called I-Net exists in many communities, but is only used in the most rarest cases. One of the reasons may be the lack of interest by city departments to interconnect. Also, distance learning does not seem to have a mission within a community, so schools do not communicate with each other. Whatever the reasons for this underutilization of the I-Nets may be, the cable companies have not promoted the empty spectrum because of their own inabilities to attract revenues.

A large number of two-way instructional TV networks exist in the campus environment of universities, colleges, federal and state facilities.

Instructional TV consists of a combination of services, i.e.

a)   the ability to order a video from a classroom and to be able manipulate the video transmission
b)   the ability to set up a bi-directional videoconference between any location on the network
c)   the ability to originate a video transmission from any location on the network and distribute it to every other network location

The capacity of instructional TV networks is quite formidable. In the forward direction, 50 to 80 channels can be transmitted simultaneously. In the return or inbound direction, 17 to 30 channels can be transmitted simultaneously.

This means that up to 30 two-way videoconference transmissions can take place at the same time. In addition, another 25 to 55 channels can be used for video-on-demand or video retrieval services.

The coaxial cable, therefore, represents a powerful tool for almost limitless expansion of interactive instructional video within a building or a campus area. Cost-effective analog transmission equipment can be used now and digital transmission can be added at any time and when it becomes available at reduced price levels.

Video Teleconferencing

Corporate Teleconferencing

During the past decade, the transmission of bi-directional video between two or more locations has grown mainly in the corporate arena to reduce travel costs and to save time.

Until very recently, videoconferencing has not been accomplished in full motion, high-quality broadcast TV, unless the transmission was between two locations on a coaxial cable system.

Codecs, to compress the video information, are available for T-I or for 772 Mbit/s (half a T-I) and for dual 56 or 64 Kbps. The transmission can be over satellites or the common carrier network.

Full-motion, compressed video requires a DS-3 band of 44.3 Mbps for high-quality transmission. Even though compression technology has been rapidly improving, it is difficult to expect that I.544 Mbps, or dual 56 or 64 Kbps, can provide true full-motion quality. And, in deed, it is not. What may be entirely sufficient quality for looking at stationary people, "talking heads" cannot be used to transmit a tennis game, any other fast-moving activity or a medical image for diagnosis.

In summary, high-quality videoconferencing can be provided in analog form on a private coaxial cable system. When connecting to the outside world quality degradations, at this present point in time, must be tolerated. In the future, true full-motion video may be provided soon utilizing new transmission and compression standards.

Distance Learning

The ability of a special teacher to communicate visually with special students has been a welcomed enhancement to the curriculum of many schools.

Very similar to the corporate teleconferencing transmission, speeds may vary between T-I, one-half T-I, dual 56 Kbit/s or ISDN. In Maryland, a school-interconnect system provides distance learning at DS-3 speeds.
The problem with distance learning systems is that portable terminals are moved to a classroom and now become a permanent fixture because the telephone company has brought in a special pair to accommodate the T-I codec.

In an enterprise, the connection to the outside world should be made at the gateway. A T-I modem on the broadband coaxial cable network can provide T-I to any classroom. As a result, the teleconferencing equipment can be moved to any classroom or lecture hall.

Telemedicine

Early telemedicine connections in the 1970s were established mostly between hospitals to further the exchange of medical information and as a training tool.

In order to achieve broadcast-quality video in both directions, the chosen media was exclusively point-to-point microwave.The complexities of establishing line-of-sight paths, the FCC filing requirements, property leases and high costs of equipment and installation have restricted the development of telemedicine during the past decade.

The development of digital and compression technology, combined with the increased need to control health care costs, will boost the growth of diagnostic services by video. In order to do so, telemedicine must become a multipoint-switched network and permit a hospital to expand its geographical area of interest through flexible growth.

To further the growth of telemedicine, true diagnostic value of the video presentation has to be assured. Within a hospital or a hospital campus, NTSC video can easily be provided over an HFC or a broadband coaxial network.

To reach outlying areas, the two-way video connection is routed through the gateway of the network and connected through common-carrier provided DS-3 or T-I circuits to the remote location. T-I or DS-3 on a SONET ring can provide ATM-switched connections to outlying medical centers. Within the hospital campus, the broadband coaxial facility can provide a switchable multipoint architecture and provide distributed telemedicine to any participating physician and medical office location.

Desktop Videoconferencing

During the last two years, the CD-ROM drive has been introduced by the computer industry to advance the computer towards sound and video. The development is usually referred to as multimedia. A number of manufacturers independently developed various playback speeds and varying number of frames per second without standardization. The result is a mediocre presentation of moving still-pictures that cannot, at this time, compete with full-motion video.

The computer industry is in the process of developing a video standard, which, even though not NTSC video, will produce smoother motion and be transmittable over the telephone network. The MPEG-I standard is described in more detail in following sections.

Desktop video and associated multipoint videoconferencing is also possible on type-5 UTP wiring at data rates of 45 Mbps or DS-3. This new technology provides a migration path to ATM-switched networks.

Capuswide desktop videoconferencing can readily be provided to NTSC full-motion video standards over the broadband coaxial network. All that is required is a VGA to NTSC card to convert the scanning rates. There are a number of videoconferencing command and control software packages available that can be used to set up multipoint desktop video connections. This software can be used as well to control the video matrix switch. Control and signalling travels on the data network while 6 MHz video is routed over the broadband coaxial network. As a result, the data network loading is not affected by high-speed video transmission requirements which otherwise would curtail the throughput of the entire network.

With all the new digital transmission standards pointing towards video in digital form, why bother with video in analog form?

The answer to this question is related to the geography of an installation and to the budgetary constraints that you may be experiencing now or in the future. To gain a better insight into the pros and cons of analog and digital video transmission methods, we have to analyze the present and future video transmission standards.

52

The NTSC Analog Video Transmission Standard

Both the NTSC and the RS-170 standard for analog transmission of video and associated audio were developed in the late 1930s.

The analog video signal is contained in a 4.2 MHz signal. The resolution or quality of the video picture is directly related to the number of horizontal lines and the number of picture updates or frames per second. NTSC transmission causes a horizontal line to be written in 63.56 microseconds and a total of 440 lines to appear to make up one frame.

A minimum of 30 frames per second are required to give the human eye the illusion of a smooth, full-motion video or film-like presentation. A standard TV set does not show us 440 lines, but more like 330 lines with a 3:4 aspect ratio.

NTSC (National Television Standards Committee) is not a global standard, but has served the broadcast community well and has promoted the growth of Television during the past decade. The resolution of the NTSC video picture provides sufficient information detail to use it for diagnostic sequences in laprascopic interventions. The resolution is sufficient for fast movements of, for instance, a football without distorting the back ground, but it is not sufficient to read typewriter-size letters on a TV set.

The computer screen in the digital VGA format has more pixels to represent finer details than NTSC video ever could do, but requires 150 Mbit/s to transmit one frame in an uncompressed format.

Disadvantages of NTSC Analog Video

The Interference Potential

NTSC analog video, when broadcasted through the air, has a number of interference problems, which are directly related to the analog transmission domain.

There is co-channel interference, which occurs when the TV-set antenna picks up two signals on the same channel frequency. Co-channel interference produces horizontal lines across a TV picture.

Co-channel interference and bad or snowy reception in over-the-horizon communities led to the inception of Community Antennas and cable TV.

The worst interference contributor, however, is multipath interference. This interference is caused by reflections of the TV signal between buildings. A TV-set antenna may receive 2,3 or more pictures, which manifests itself on the TV set as "ghosting".

Ghosting in downtown and urban areas led to the development of the set-top converter, which can bring all channels to the TV set on Channel 3.

Cable-TV systems transmitting TV channels in adjacent 6 MHz bands added interference products of their own. Adjacent channel-filtering, the number of cascaded amplifiers along the distribution system as well as ingress and egress at loose connector locations, added "herringbones", "windshield wipers" and "hum bars" to co-channel and ghosting.

In defense of multichannel analog video transmission, it should be noted that in-building and in-campus systems, i.e. systems of limited sizes, can be built without concern for interference or picture degradation and constitute the most cost-effective method of video transportation, whether in analog or digital format. Proof of this lies in the wisdom that new voice/data delivery systems will use coaxial cable for the last mile.

Regional Standardization

Analog video transmission systems were standardized during a time period that did not foresee any global uses of video.

There are three major analog transmission standards: NTSC in the U.S., Canada and Mexico, PAL and SECAM, which are used throughout Europe, Asia and South America.

Both PAL and SECAM utilize a wider frequency spectrum than NTSC. Therefore, definition and resolution values are higher, but different from each other. The translation of transmission standards at border points between Germany and France, for instance, has been time-consuming and costly. Both microwave and satellite transmissions are affected. Our shrinking globe demands new international standards.

The Many Advantages of Analog Video

Despite the much heralded arrival and growth of digital video transmission, there are many advantages to enterprise networks employing proven and economical analog video transmission.

- A single broadband coaxial cable can transmit over lOO simultaneous 6 MHz video transmissions

- The cable equipment components such as amplifiers, power supplies, and passive components such as splitters, muiltitaps and connectors, are inexpensive. Supply and demand for multichannel video system components have kept component pricing at levels of the 1960s

- The costs of TV sets have dropped over the years and clarity, features and luminance have increased

- The cable-ready TV set can tune in 100 channels by pushbutton control. The improved tuners accommodate adjacent channel reception without any interference

- In an enterprise network, multichannel video reception can be provided at lowest expense by just connecting the TV set to the cable and without the need of a set-top connector

- Analog video can readily be imported to any computer screen using inexpensive scan converter cards

The New Digital Video Standards

MPEG-2 Broadcast Quality Television

The Motion Picture Experts Group (MPEG) has developed new digital transmission standards during the past several years and in conjunction with international standard organizations.

For the first time in the history of television, a single international and global standard is being adopted for the transmission of broadcast quality full-motion video. The application of the new digital transmission standard applies to all video-delivery methods such as broadcast TV, cable TV, videoconferencing, video-on-demand, satellite up and down links as well as the transmission on the information superhighway.

The development of the MPEG-2 standard is directly related to the quantum-leap in digital compression technology. Originally developed for teleconferencing of NTSC video over satellites, the compression technology was further refined to accommodate DBS (Direct Broadcast Satellite) services and now forms the basis of any digital video transmission.

Digital compression technology is composed of a step-by-step procedure to eliminate unnecessary data. While compression of analog information is detrimental to the signal quality, digital compression permits a 500% decrease of bandwidth. While a 6 MHz channel is required for the transmission of an analog video channel, a total of 5 compressed, digital video transmissions can be accommodated in the same 6 MHz allocation.

This changes the economics of television for broadcast stations, satellite and cable TV systems. The hype about cable TV distribution systems carrying 500 channels is directly related to digital compression technology. What is being talked about is a bandwidth capacity of one hundred 6 MHz channels in a coaxial cable network operating between 54 and 750 MHz.

It should be noted, however, that digital TV transmission ends at the set-top converter in front of the TV set. As long as the price of a digital TV set does not equal that of our present-day analog TV set, digital compression and conversion to analog NTSC will be the convention.

It is noted that private or campuswide video networks on coaxial cable or fiber will retain analog NTSC transmission standards for a long time to avoid costly digital compression and set-top converter decompression technology.

MPEG-2, which is also the standard of the International Standards Organization (ISO 13818), is not only a compression standard, but also a transmission standard, which makes it a blueprint for system interoperability.

The MPEG-2 Compression Standard

The MPEG-2 standard defines the delivery of a full-screen resolution of 704 x 480 pixels at 60 interlaced fields per second.

An NTSC broadcast analog video transmission requires a 6 MHz spectrum. When digitized, the partially compressed video transmission requires a minimum of a 44.5 Mbps transmission rate (DS-3).

MPEG-2 defines that 5 video channels shall be accommodated in a 6 MHz spectrum. Therefore, the transmission rate in digital form cannot exceed 9 Mbps.

This standard then relies on formidable compression techniques to accomplish this reduction in the rate of transmission. This data-elimination process involves a step-by-step procedure. Each step in the procedure focuses on a specific element of the video signal and eliminates unnecessary data within that area. The efficiency of each of these data-elimination functions has to be provided while retaining the highest possible quality of the picture presentation.

Some of the more common data-elimination methods are:

a)   elimination of background detail when the camera pans rapidly

b)   the transmitted data is minimized by motion compensation - taking advantage of similarities between adjacent frames

c)   discrete cosine transform (DCT) function minimizes spatial redundancy between adjacent pixels within a single frame

d)   minimizing the quantity of data by quantization of continuous range numbers to discrete, predetermined value

e)   further reduction of data using a lossless coding technique called eutrophy coding before transmission

It is recognized that the encoding process has to be decoded at the TV set. Cable TV companies propose to do all decoding functions in the set-top converter.

It will take a long time until digital TV sets with decoding ability are available to the consumer at present TV-set price levels.

The MPEG-2 Transmission Standard

The primary mode of digital video transmission in broadcast applications will be the transport stream (TS). The transport stream utilizes fixed-length packets of relatively short length (188 bytes). The transport stream is specifically designed for enhanced error resiliency and packet-loss detection.

Many programs, i.e. video, audio, data can be combined in one transport stream. This means that terrestrial broadcast, cable TV, satellites, data download and interactive telephony-based services are only a few of the emerging applications of the transport stream.

The common structure of the TS is the packetized elementary stream (PES). All MPEG video and audio data must be formatted into a PES and inserted into the payload portion of the TS packet.

MPEG-2 adopts a hierarchical view of transmission systems. A network may have one or more transport streams. A transport stream may contain one or more programs and a program may contain one or more elementary streams.

One of the main advantages of packet-based transport is the straightforward support of drop/add and insert of asynchronous services. This permits the distribution of services via high-rate services such as SONET OC-3 with later partitioning to lower-rate transports such as a 6 MHz video channel utilizing

44.5 Mbps.

The MPEG-2 is a network and data-link independent transport and has excellent affinity to the ATM-layered approach. The length of the MPEG-TS transport packet was chosen to map evenly into four ATM-cell payloads. The MPEG-2 system specification provides support for clock-recovery, synchronization, buffer management, private data and conditional access. It can become a major foundation of interoperability in the "digital information superhighway".

The Advantages of MPEG-2 Transmission

It is forecasted that the market will be choking on MPEG-2 boards with supply outstripping the demand already by no later than 1996.

- MPEG-2 is a global standard and will become the choice in international video transmission systems

- MPEG-2 can be ATM-switched and packetized into a SONET ring-architecture for long-distance transmissions

- MPEG-2 is readily converted to NTSC and then transmitted on a broadband coaxial network in a 1.5 MHz frequency spectrum

- MPEG-2 boards will be plentiful and subject to price erosion over a relatively short time period

- MPEG-2 will be the video format of choice for video transmissions to the outside world such as regional interconnections, distance learning, satellite teleconferencing, near video-on-demand (NVOD) and video-on-demand (VOD) transmissions to the home and to and from enterprise video networks

- A computer equipped with an MPEG-2 board can be connected to the coaxial drop cable of the serving cable company via a cable modem. The cable modem simply provides a spectrum slot for the video transmission to the headend. Conversion to NTSC for analog downstream transmission to inexpensive TV sets within a campus can be achieved using a scan converter

- The combination scan converter/cable modem can transfer MPEG-2 to NTSC for analog transmission on the broadband coaxial network. Circuit-switching at the headend accommodates multipoint desktop videoconferencing with participation of TV set and projection equipment

- The MPEG-2 specification will not only provide fast data rates, but also assure highest picture quality. MPEG-2 delivers full-screen resolution of 704 x 480 pixels. This high-resolution specification is known as the CCIR 601 format, the global standard from the Communications Committee International for Radio. With 30 frames or 60 interlaced fields per second, full-motion, film-like video is provided with a resolution that exceeds that of the NTSC analog standard

The Disadvantages of MPEG-2 Transmission

MPEG-2 requires compression and decompression equipment at either end of the circuit. The need for such equipment and the 5-8 Mbit/s speed requirement demand costly upgrading of the existing data networks.

- MEG-2 cannot be decoded by a standard television set. Special set-top converter/decoder units are required

- Multichannel video networking using MEG-2 video requires high-speed digital networking throughout the enterprise. Even fast ethernet or FDDI may not provide sufficient throughput at a 100 Mbit/s rate to permit video growth without excessive overloading

Advanced Television (ATV) or High-Definition Television (HDTV)

HDTV has been in the news for a few years. Early japanese formats were not successful in the U.S. because the FCC decided that new digital transmission methods must interface with existing formats and cannot obsolete the present TV set population.

MPEG-3 was originally supposed to be developed to accommodate HDTV until it was determined that MPEG-2 could be extended to include HDTV capability.

While the MPEG-2 standard permits full-motion video at a digital speed of 6-9 Mbps, future HDTV standards require about 20 Mbps. The DS-3 standard of 45 Mbps is an integral part of the AMT (Asynchronous Transfer Mode) architecture and will accommodate either 5 simultaneous MPEG-2 full-motion video channels or a minimum of 2 HDTV video channels with 1080 active interlaced lines.

This also means that either 5 standard, full-motion digital video channels or 2 HDTV channels can be transported in a conventional 6 MHz band on a coaxial broadband network or in over-the-air broadcast applications.

The HDTV standardization is not finalized at this time. The "Grand Alliance", a consortium of seven companies, is conducting further tests at the Advanced Television Test Center (ATTC) in Alexandria, VA.

The open issues are:

a) the selection of the winning modulation system. The present choices are quadrature amplitude modulation (QAM), vestigial sideband modulation (VSM) or the european-developed (COFDM) coherent orthoginal modulation system, which provides better rejection of multipath problems

b) the decision on the argument that every HDTV set also has to be a computer monitor. TV sets use an interlace-scan option. Computer monitors use progressive scanning, which reduces the "flicker" associated with interlace. However, interlace promises a slightly better picture quality and lower priced TV sets

It is expected that the Advisory Committee will propose an agreed upon standard to the FCC in early 1995, which may not coincide with the standardization process in Europe.

Even though, the establishment of HDTV promises to provide added flexibility and services to the broadcast industry, the progress has been slowed because of uncompleted standards, high initial investment (about 1 million per transmitter), lack of combination analog/digital television set manufacture.

Even though, the 9x16 aspect ratio is very pleasing to the eye in combination with the high resolution that 1080 lines can provide, HDTV cannot be expected to be established in a short time frame. Similar to the conversion from black and white to color, which took a decade, it is expected that a conversion to HDTV may take two decades and may even be restricted to the average user for many years because of high TV set prices.

The MPEG-I Standard

The Motion Picture Experts Group has set a minimum standard for display of images. The MPEG-I standard calls for 320x240 pixels at 15 frames per second. Since the human eye needs to see a minimum of 30 frames per second to have the appearance of continuous motion, MPEG-I provides for 30 frames per second in a quarter of the screen area of a computer display.

In developing the MPEG-I standard, a compromise was reached between appearance and transmission. Even though MPEG-I has a somewhat jerky appearance, the video is compressed to utilize transmission rates of about I5O Kbps. This transmission speed is 4O times less than the 6 Mbps rate of the MPEG-2 standard for full-motion video.

The MPEG-I standard may serve well in videoconference settings that utilize ethernet LANs and that may be connected to the outside world via ISDN-switched network.

While it uses the discrete cosine compression technique that JPEG uses, MPEG-I improves on JPEG's compression ratios by reducing frame-to-frame redundancies. As in animated cartoons, backgrounds are often stable over many frames, while objects move against them.

MPEG-I exploits this fact in achieving its 2OO:I ratios, compared to JPEG's 4O:I. Right now, a CD-ROM can hold 74 minutes of footage, thanks to MPEG-I.

Most MPEG-I compression is presently done by high-dollar service bureaus that have invested in the necessary computing power. Even these take several times longer to compress a piece than the actual run-time of the footage; with desktop implementations, compressing a three-minute music video might take half a day.

MPEG-I lacks MPEG-2's field/frame coding capability, which delivers better quality, especially in high-speed sequences such as sports events. It is unlikely that MPEG-I systems will last a long time in interactive TV. MPEG-I, however, may play a transitional role and may be moving up to MPEG-I.5, a compromise intended for digital ad insertion and movies.

Fig. 3-2 shows the various MPEG compression approaches in relation to their data rate requirements and expected uses. Whatever the future brings, the HFC broadband cable will be ready to carry any MPEG data rate in a fraction of a 6 MHz channel presently used for one analog video transmission.

Desktop Video Standards

The most volatile conditions exist in the desktop industry since the introduction of CD-ROM drives and the concept of multimedia presentations on personal computers.

Many video compression algorithms have been developed for variety of playback sizes and frames per second. The resulting pictures have good resolution but feature jerky motion because of playback rates below 3O frames per second.

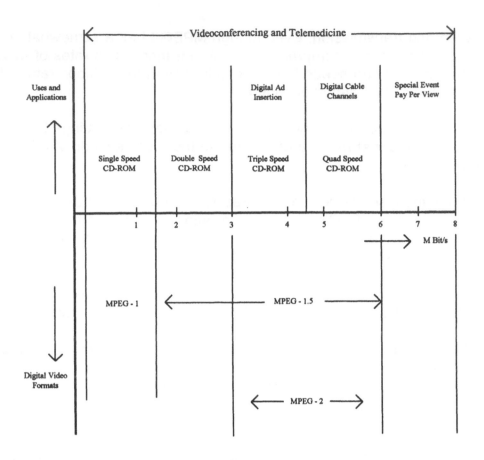

Fig. 3-2 Digital Video Formats,
Uses and Applications

The various multimedia presentations presently being sold may vary from I6OxI2O pixels at I5 frames per second to higher pixel numbers at 3O frames per second.

The desire to launch CD-ROM products was greater than the ability to wait for completed standards. As a result, there are resolutions down to I6OxI2O pixels and I5 frames per second. Bad resolution, combined with jerky pictures, is not expected to be the wave of the future in multimedia. Obviously, CD-ROM games will sell well despite the lack of resolution and jerky motion.

However, these inferior systems are not usable for desktop videoconferencing. Before purchasing a CD-ROM multimedia computer, the buyer is well advised to inquire as to resolution, number of pixels and frame rate.

Reflections

Video transmission in a digital format is on the drawing board and is going through changes and quality improvements. With the advent of NVOD and VOD (Video-on- Demand), the use of set-top converters will multiply.

The set-top converter decodes the compressed digital signal back to analog video for consumption by the TV set. At that point in time, it is expected that the prices of digital to analog conversion equipment will fall to reasonable levels.

While MPEG-I is moving up in data rates, MPEG-2 may be operating with reduced data rates. The critical eye of the user will determine when image degradation becomes noticeable.

Fig. 3-2 indicates the data rates of the various digital video transmission formats together with the expected applications and uses. If image degradation becomes noticeable between 3 Mbit/s and 5 Mbit/s, the market will move towards MPEG-2 as the only full-motion digital video format. If MPEG-2 establishes itself in the range of 3-8 Mbit/s, it is possible to carry 5 to IO digital video channels in one 6 MHz frequency assignment on the broadband HFC cable network.

What do these digital video formats mean to the corporate video network?

- A broadband fiber coaxial network will carry them all with ease

- A data LAN, even at FDDI or SONET data rates, cannot support large numbers of internal video communications occurring at the same time

- Video transmission from a computer terminal can travel in digital form on the data network or on the broadband network

- On the data network, a data rate of 8 Mbit/s for one video transmission can become a large percentage of the network throughput (8 percent of a IOO Mbit/s LAN)

- On a broadband coaxial network, the same 8 Mbit/s occupy 2O% of one 6 MHz analog video channel. With over 3O two-way channels available, the digital single-channel video transmission only represents O.6 percent of the total two-way capacity.

Should digital video transmission be a high priority in the establishment of a campuswide video network?

- The first priority for an enterprise must be the establishment of a high-capacity video transmission network. Whether the video is transmitted in analog or digital form is a decision that can be made whenever one form of transmission is more economical than the other

- At the present time, video transmission in analog form must be considered more economical. The reasons for this cost-effectiveness are linked to proven cable-television components, inexpensive television sets and quality full-motion video presentation.

Why is the broadband coaxial cable network not frequently used by corporations and enterprises at the present time?

- There are many corporations and also large government facilities that use broadband coaxial cable networks for high-speed Local Area Networks. Air Force bases use dual-cable 50-450 MHz facilities for almost a decade

- Broadband LANs are used in robotic manufacturing facilities as well as GBit/s facilities for multiple and channelized high-speed data transfer

- Two-way broadband communication is just now coming into use for training, educational aid, telemedicine and videoconferencing

- There are no products to be sold until the broadband system is in place. Therefore, the vendors feel that it is an easier sell to promote products for twisted-pair and fiber

How can I assure full-motion videoconferencing now and avoid costly upgrading of the data network?

- By designing and installing a broadband coaxial or HFC network throughout your enterprise and by using NTSC full-motion video between all participant locations

- Analog NTSC full-motion video will be around for a long time, as it can utilize standard and inexpensive TV sets for direct viewing. A number of companies are producing transfer boards that permit the transfer of NTSC video to the VGA computer monitor. Likewise, it is possible to transfer the image on the computer screen together with the picture of a camera to NTSC analog video for modulation on the broadband coaxial network

- The networking of this analog-based high-quality NTSC video can be provided today and at reasonable prices

    - by connecting the workstation to a two-way coaxial cable network (video/audio) and using the established LAN for command and control

- by using the proprietary videoconferencing systems, which require type 3 or type 5 (four pairs) coaxial or fiber cable to connect to any participating workstation or location

- Enterprisewide NTSC analog video transmission is recommended for the development of videoconferencing and telemedicine networks. Migration towards ATM-based digital systems then becomes a logical path for video traffic entering and exiting the enterprise network. Such applications may include distance learning, teleconferencing and telemedicine networks to remote locations.

Present and Future Public Network Long Distance Standards

ISDN (Integrated Services Digital Network) and Switched 56 Kbit/s

There is a massive movement towards videoconferencing products that connect the desktop with the outside world. To enable these products to talk to each other, standards have been developed that describe how videoconferencing units compress and transmit video and audio information over common carrier public networks such as ISDN and switched 56 Kbit/s lines. The H.320 standard is the umbrella for the video compression standard H.261 (Px64) and several audio compression standards such as G.711, G.722 and G.728. G.711 covers baseband audio from 50-3600 Hz at 48-64 Kbps, G.722 covers 50-6900 Hz audio at 48-64 Kbps and 6.728 covers 50-3600 Hz audio at 16 Kbps. A data-sharing capability between desktop videoconferencing systems is being developed under T.120.
Units complying to these standards can communicate with each other.

The 56 Kbit/s-switched service has been promoted for "fast" data transfer over the public telephone network. Switched 56 Kbit/s service works over standard copper wires and was established before any upgrade to fiber was contemplated by the RBOCs.

The signal level of the switched 56/Kbit/s service is called DS 0. The DS designation stands for Digital Signal Zero, which covers a bit rate of 64 Kbit/s and can transport one voice channel.

Also operating in the DS 0 signal level is ISDN (Integrated Services Digital Network). The idea behind ISDN is to provide a common interface standard through a single jack in the wall through which all separate transmission services for voice, data and imaging are integrated.

Voice, data and images can arrive simultaneously over the same pair of copper wires that are already in place. ISDN is a switched service, switched through the central offices of the local telephone companies or through the switches of the long-distance carriers.

There are two ISDN transmission rates. With Basic Rate Interface (BRI), the local telephone company divides its twisted-pair local loop into three separate channels: two 64 Kbps B-channels and one I6 Kbps D-channel. The B-channels are used to send the voice or data and the D-channel carries signaling and other data for controlling the call.

Primary Rate Interface (PRI) consists of twenty-three 64 Kbps B-channels and one 64 Kbps D-channel. With the PRI, information can be sent at TI speeds using 4 wires.

It is the operation of the D-channel that distinguishes ISDN from other digital alternatives to the analog network. Signalling in the D-channel tells the network how to handle the B-channel data. It also makes possible supplementary services and provides the user with call-control information. Many of the new computer telephony applications can take advantage of ISDN circuit capabilities.

It is obvious that a 64 Kbit/s speed cannot be used for real-time video. Full-motion video cannot be transmitted over ISDN wires. However, switched 56 Kbit/s and ISDN can be used for still pictures that, if you have time, you can see develop on your computer screen.

Switched 56 Kbit/s and ISDN are being exploited by a number of vendors for new desktop conferencing products. Desktop conferencing does not necessarily require video. What is being accomplished quite well on ISDN is audioconferencing and document sharing. Talking-heads video transmission is not adding much to a discussion about a document and in the process of what should be changed in this document.

It is interesting to note that desktop-conferencing software with minor modifications can be made available for videoconferencing applications and used on higher speed transmission facilities.

T-I Networking

Commonly referred to as DS I, the transmission speed of a T-I link is I.544 Mbit/s. This speed can accommodate 24 voice grade 64 Kbit/s transmissions simultaneously.

On a T-I, signals are multiplexed so that two pairs of wire can carry 24 separate voice or data conversations. These same two pairs of wire would carry only two voice conversations using analog technology. With each conversation, 64 Kbps and another 8 Kbps being used for control signals, the total T-I capacity is I.544 Mbps (million bits per second).

T-I has been used to connect a company directly to its long-distance carrier or to the local telephone company central office for many years. This is often presented at a cost saving to the user. The greater cost saving, however, is to the carrier, requiring only 4 wires to deliver 24 circuits.

Many companies also connect their locations for voice and data communications with T-I circuits. You may also hear the term "Fractional T-I", which means that you are getting a circuit with some fraction or increment of the total T-I capacity. For example, the circuit may provide only twelve 64 Kbps channels. But if we analyze the T-I connection, we find either 24 voice channels or I6O computer ports that can be running at 96OO bit/s each and simultaneously.

The following table shows the capabilities of a T-I circuit:

Signal Level	Carrier	Rate	Voice Circuits	64 Kbit/s Data	Video
DS-I	T-I	I.544 Mbit/s	24	24	I

It is noted that the transmission of video over a T-I connection requires expensive compression equipment to provide a real-time, full-motion video with fairly good resolution.

T-I can share between different digital services and between voice and data. The T-I multiplexer can make the usage assignment. For instance, when IO voice lines are required, the remaining bandwidth is 896 Kbit/s. This remaining spectrum can be used for fourteen 64 Kbit/s data circuits or ten 64 Kbit/s data connections plus twenty-six slow 96OO bit/sec data streams.

Voice compression equipment can increase the number of voice channels to two per 64 Kbit/s segment for at least 44 channels.

What are the Advantages of T-I?

- A T-I connection consists of two wire pairs. While in the past each of the voice connections required a single twisted-pair, T-I reduces the cable requirement from 24 to 2

- The T-I connection can transmit voice, data and compressed video signals in digital form and at varying speeds

- Using special compression equipment, the transmission of video is possible with fairly good resolution, but substandard to the NTSC analog standard

What are the Disadvantages of T-I?

- One of the problems of T-I transmission is the requirement of good twisted-pairs and for regeneration equipment every 6OOO ft. Copper pairs are composites of different gauge copper wire and have to be selected, tested and controlled

- A T-I connection may be routed through a number of central offices and not at all take a direct route. Yet, the pricing will be based on the actual route within the local loop

- While fractures of a T-I circuit can be used for voice and data, full-motion video transmission must be considered marginal since I.5 Mbit/s do not compare favorable with the requirement of 5 Mbit/s for high-quality resolution

T-I C Networking

By placing regenerators at about 3OOO ft. intervals, the data rate of a T-I network can be doubled.

Signal Level	Carrier	Rate	Voice Circuits	64 Kbit/s Data	Video
DS-I C	T-I C	3.I52 Mbit/s	48	48	I or 2

T-I C is a good choice for video transmission to a remote location such as an associated medical facility of a hospital for telemedicine and diagnostic evaluation. As long as the facility is intra LATA, costs may be reasonable and comparable to T-I.

The Digital Transmission Hierarchy

Since we are seeing increasing data rates and falling costs in the public telephone network, it is appropriate to take a look at other evolving data rates. The table compares north-american and CCITT global transmission formats:

U.S or North American Formats

Signal Level Carrier	Bit Rate	Circuits or Voice Channels	64kbps Data	Video Ch 45 Mbps	Video Ch MPEG - 2
DS - 0 / —	64 kbps	1	1	—	—
DS -1-/ T-1	1.544 Mbps	24	24	—	1
DS-1C/T-1C	3.152 Mbps	48	48	—	1
DS -2 / T-2	6.312 Mbps	96	96	—	1
DS -3 / T-3	44.736 Mbps	672	672	1	7
DS -3A/T-3A	89.472 Mbps	1,344	1,344	2	14
DS - 4 /T-4	274.17Mbps	4,032	4032	12	45
SONET Signal Level					
OC - 1	54.00 Mbps	810	810	1	8
OC - 2	155.52Mbps	2430	2430	3	18
OC - 3	466.56Mbps	7290	7290	9	54
OC - 4	622.08Mbps	9720	9720	12	96

International Formats

CCITT Signal Level	Bit Rate	Circuits or Voice Channels	64 kbps Data	Video Ch 45 Mbps	Video Ch MPEG - 2
0	64 kbps	1	1	—	—
1 / E-1	2.048 Mbps	30	30	—	—
2 / E-2	8.448 Mbps	120	120	—	1
3 / E-3	34.368 Mbps	480	480	1	5
4 / E-4	139.26Mbps	1920	1920	2	23
5 / E-5	565.14Mbps	7680	7680	10	92
CCITT Signal Level					
1 / SDH - 1	155.52 Mbps	2430	2430	3	18
4 / SDH - 4	622.08 Mbps	9720	9720	12	96

Table 3 - 1 : The Digital Transmission Hierarchy

Global Standards

A comparison between the north american and international formats shows that the highest data rates are uniform. This makes global data transfer possible without requiring any interface equipment.

SDH-I and OC-2 are at I55.52 Mbit/s - SDH-4 and OC-4 are at 622.O8 Mbit/s.

Video Transmission

While we were talking about a video channel on a T-I facility, the transmission of multichannel, bi-directional video channels is restricted to the formats with high data rates. The SONET OC-I to OC-4 hierarchy can do that, but only OC-3 and OC-4 equipment will exceed the channel capacity of a broadband coaxial cable. It is obvious that these high-speed formats require expensive processing equipment and command highest costs, as they are developed for global video transfer.

Beware of the vendor that promotes a SONET ring with ATM switching for your data network upgrade and tells you that his product will handle video. Yes, he did not lie, SONET with ATM switching will handle video, but the number of simultaneous video transmissions, for instance, in an OC-I system is <u>one</u>, or with expensive MPEG-2 compression is <u>eight</u>.

While digital video transmission to the outside world is the standard, enterprisewide bi-directional, full-motion video communication requires the capacity of a broadband coaxial network to grow unrestricted and without the cost of expensive packet-switching equipment. As a byproduct, the transmission of video in analog form will conserve costs and make use of inexpensive TV sets and tuners.

HFC Transmission

The cable and telephone companies will use the international Signal Level I E-I carrier for voice and data on the new hybrid fiber/coaxial broadband system (HFC).

3O voice channels in the E-I data rate will be transmitted using (AM) amplitude-modulation on the fiber and distribution coax to the home. The return path of the voice channel is digitized and then amplitude-modulated to the coax cable, merged into the fiber network and processed at the headend.

Comparing the HFC transmission of the cable and telephone companies with the enterprise campus HFC network, one concludes that both are identical except for much shorter distances in a campus. On that basis you can do as well or better, i.e. the same E-I 3O-channel-grouped voice transmission on coax and fiber can be brought to the headend, gateway or video operation center for transfer to the public network.

The DS-3/T-3 Standard

At 44.736 Mbit/s, the DS-3/T-3 standard permits full-motion video transmission without expensive compression technology. DS-3/T-3 is the data rate that interfaces with OC-I.

The DS-3 standard assumes (ATM) asynchronous transfer mode interoperability at 45 Mbps.

A recent ad campaign by AT&T promotes a videoconference package called Enhanced Multimedia Interface (EMMI), which can bring 45 Mbps full-motion video to the desktop for under $ 8,000. Other companies are developing APC-DS3 insert boards for PCs that will deliver T-3 image and data transfer speeds of 45 Mbps.

APC-DS3 maximizes data transmission with minimal demand on the host CPU, freeing the host CPU to perform the other tasks for which it is best suited. APC-DS3 applications utilize the same software architecture as ATM and are, therefore, interoperable with ATM-based networks.

The DS-3 standard and the evolving 45 Mbps video transmission system can establish videoconferencing between any PC connected to a fast ethernet network and an ATM switch at high prices. Using scan conversion to NTSC at the PC, and using the broadband cable for enterprisewide transmission, appears more cost-effective.
The DS-3/T-3 format can then be established at the gateway for traffic to the outside world.

SONET and ATM Switching

SONET (Synchronous Optical Network) requires single-mode fiber-optic cables as the transmission medium. Only the fiber-optic cable assures the ability to increase the data rate and migrate from OC-I to OC-4 transmission. Every digital transmission consist of ones and zeros.

Framing

To decipher a bit stream, the sending unit must insert reference marks and the receiving unit must be able to read the reference marks. This is called framing. Framing is not payload and, therefore, called overhead. Time Division Multiplexing (TDM) is a commonly used framing method that can be used to divide the bit stream into separate channels and permits the channel always to occupy the same position in each frame.

Standard T-I is time division multiplexing. Every I93rd bit in a standard DS-I signal is a framing bit. The I.544 Mbit/s, in actuality, exist of 24x64,000 bit/s, which equals I.536 Mbit/s plus 8000 bit/s overhead for framing.

SONET framing has more overhead and more payload. The framing units in SONET are bytes or octets rather than bits. SONET framing sets aside 4% of the total bandwidth for overhead. This extra bandwidth can support multiple supervisory order wires and nested addresses for numerous protocol layers.

The Synchronous Transport Signal (STS)

STS, the synchronous transport signal, is the electrical version of the SONET signal, which is defined only on optical fiber. The SONET optical carrier level one (OC-I) is the direct translation of the STS-I electrical signal into an optical signal. Their information and overhead contents are the same.

STS is byte-oriented. The digital unit within STS is an octet (8 bits). Octet may sound the same as a byte, but "byte" implies a character or other unit that has meaning by itself, and bytes sometimes are more than 8 bits. The term octet is used to indicate an eight-bit unit, even if it has no logical meaning on its own, like 8 bits from a graphic image file.

SONET and STS employ basic framing to perform the same function: to make whatever signal the user sends understandable to the receiver. At constant intervals (every I3.888... s) the STS transmitter inserts a block of overhead bytes. These framing blocks establish markers that allow the receiver to identify individual channels. In addition to synchronization, framing overhead includes error-checking, alarms and data channel for network management.

Fig. 3-3 shows the STS framing overhead and the 9 rows of payload in a I25 microsecond frame. The STS frame contains 9 cycles of I3.888 microseconds. Since the frame duration is the same as the T-I frame, any kind of data traffic can be transmitted over the SONET system.

Table 3 - 2 shows payloads and overheads expressed in rows. Each frame has 9 rows.

SONET Payloads and Overheads

Level OC/STS-n	Data Rate	Payload Size (octets)	Overhead (bytes)	Length of Rows (octets)
1	51.84Mbps	87	3	90
3	155.52Mbps	261	9	270
9	466.56Mbps	783	27	810
12	622.08Mbps	1,044	36	1,080
18	933.12Mbps	1,566	54	1,620
24	1244.16Mbps	2,088	72	2,160
36	1866.24Mbps	3,132	108	3,240
48	2488.32Mbps	4,176	144	4,320

Table 3 - 2 : SONET Octets

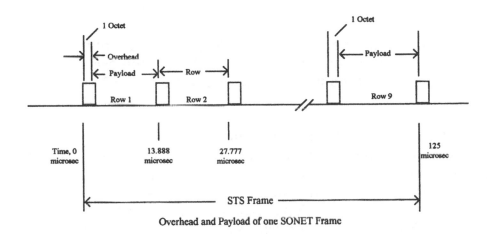

Overhead and Payload of one SONET Frame

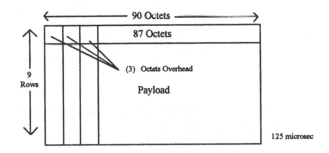

Fig. 3-3 The STS Frame with 90 Octects by 9 Rows

SONET is sometimes associated with packets, particularly frame and cell relay services. Yet, SONET is really a very fundamental framing format that can carry any kind of traffic. SONET is an intelligent time code.

Digital services continue to expand and with them the networks that support them. SONET overhead channels will allow carriers to keep up with customer demand. Without some kind of automation, even the initial installation of connections or provisioning of services would develop long lead times. Moves and changes would be painful without the management channels that permit reconfigurations at any SONET device in the network.

Fig. 3-3 shows the relationship of payload and transport overhead for each row. Every SONET frame contains 9 rows. Each OC level is a direct multiple of 5l.84 Mbit/s.

ATM Switching

ATM or asynchronous transfer mode superimposes an asynchronous handling of data streams onto the SONET synchronous transmission system.

On a physical connection between ATM nodes, the bit rate of the data stream must be synchronized with the master clock of the switch or multiplexer at one end of the transmission line. On a SONET line, the clocking is certainly synchronous, by definition, with the network's clock.

By contrast, the common asynchronous data of a communication port or terminal is sent on its own, controlled by the internal clock of the sender. The receiver is assumed to know what the bit rate is and have its own clock that is running close enough to the sender's rate so it is unnecessary to send the clock signal explicitly. Only data bits are sent between nodes. The start and stop bits in asynchronous characters allow the receiver to recognize the start of a character.

Also, the flow of cells is constant, one after the other with no spaces between them, and can be considered synchronous. Idle, or empty, cells are inserted when there is no "live data" to send so there is no interruption in the flow of cells. Consequently, the rate of cell arrival is constant (synchronous) at l/424 of the clock rate and synchronized to the same transmission or switching system clock.

The term asynchronous transfer refers to the indeterminate time when the next information may start to transmit.

The next cell in the physical flow could be from any logical connection of any user. The preceding cell predicts nothing about the next cell except when it will start and that it will start - immediately after the current cell ends. The next cell could be empty or destined for any location on any logical connection. Thus, the cells for any given connection arrive asynchronously.

74

The entire ATM system uses addresses to define how cells pass through the network. It is in addressing that switches create and maintain connections. As in frame relay, the cell address has only local significance. Any switch may change the address as a cell passes through.

An ATM switch then is a device that accepts a cell on any port, reads its address and uses the address to decide where to forward ("relay" or "transfer") the cell. Typically, the cell will be sent out to another port on a link to another node. Some connections may terminate within the node either as user traffic or for node management.

Fig. 3-4 shows the structure of the ATM cell, which consists of 53 bytes. The 5-byte header controls the communication to the distant ATM device. The remainder 48 bytes is the payload. Cell addresses prescribe the routing of a particular connection. The ATM switch forwards or transfers the cell. If there is an error, the ATM switch will discard the cell. The ATM switch assigns port numbers and cell addresses in the sequence of the arriving cells.

The fundamental building block of all ATM end-to-end connections is a one-way, single-hop connection between adjacent ATM devices called an

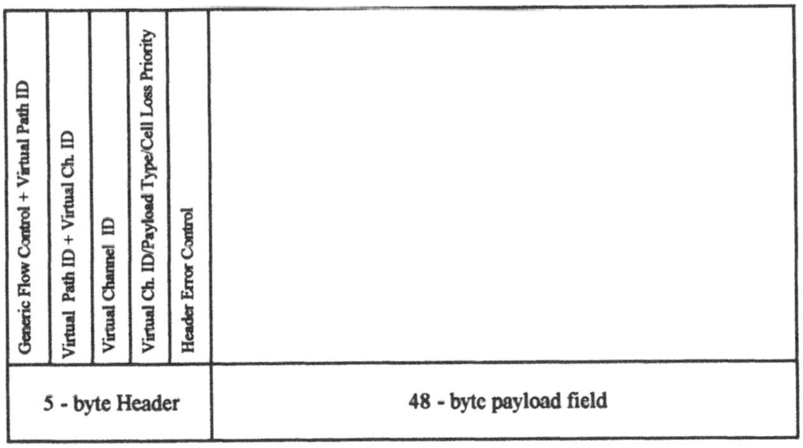

Generic Flow Control + Virtual Path ID	Virtual Path ID + Virtual Ch. ID	Virtual Channel ID	Virtual Ch. ID/Payload Type/Cell Loss Priority	Header Error Control	
5 - byte Header					48 - byte payload field

Fig. 3-4 The ATM Cell

ATM peer-to-peer (APP) connection. A physical layer connection, or channel, between devices like switches, multiplexers, bridges, routers or other equipment can carry many of these APPs.

APPs are identified by cell address, both path and connection. Cells that belong to an APP have the same address as they pass any given point in the network, though the address may be different at different points. A full-duplex connection requires two APPs, one in each direction. One can assign the same address to both directions of a full-duplex connection.

The SONET Ring Architecture

ATM switching is required only when various types of digital traffic or data rates need to be combined in a SONET transmission system. ATM is connection oriented. A path must be defined through the network before any data can be transmitted.

If we deploy ATM switches in major buildings of a campus area, we can interconnect all switch locations with a bi-directional SONET ring. The ring architecture provides survivable communication since all data streams are transmitted in both directions. Route diversity is, of course, desirable. If a different routing of the two directions cannot be obtained, it is recommended to at least use separate fiber cables in separate ducts to eliminate total outages.

Fig. 3-5 shows a SONET ring architecture strictly used to interconnect cable television headends. No ATM switch is required in this architecture since the 96 analog video channels are going to all other locations.

Using MPEG-2 standards, the number of video channels could be increased to over 300 channels. However, if one would add digital video channels in the MPEG-2 format or data and voice, ATM switches would be required.

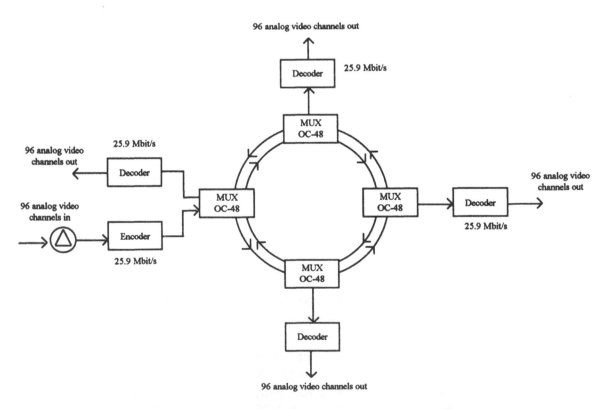

Fig. 3-5 An example of a SONET Ring for Cable Headend Interconnection

The SONET ring configuration, as shown in Fig. 3-5, requires only two fibers to provide both service and protection capabilities for the same 96 video channels. In addition, the SONET ring offers the flexibility to pick up and drop off signals at any hub site on the ring. A local channel generated from hub A, for example, can enter the SONET ring and return to the headend in the same manner that the headend delivers signals to the hub. SONET's overhead transport capability controls the transport and delivery of the digital signals to any location on the ring. SONET's two-way ring transport systems can survive network failures without any loss of services.

The example of Fig. 3-5 is recommended for enterprise multicampus interconnection of voice/data and video. Each campus would have a gateway with HFC architecture within each campus. While intra-campus video transmission is analog, campus interconnect is digital to overcome the greater distances between the locations.

When campus HFC systems go digital, using MPEG-2 compression, ATM switches are added to transfer the cell packets directly into the SONET payload structure.

SONET's "open" architecture provides a highly flexible platform that can incrementally grow as greater capacity is required.

SONET transports signals at the optical carrier rates of OC-3 (six video channels), OC-l2 (24 video channels) and OC-48 (96 video channels). Depending on the needs of an enterprise system, the SONET network could begin with an implementation of a lower capacity system that could incrementally grow as network demands and new revenues are realized.

For example, corporations can initially implement an OC-l2 SONET system in a point-to-point implementation that links two network centers together - while putting the network in place that can provide for emerging interactive and multimedia service as they become available. In the future, that same point-to-point solution can migrate to a ring architecture when additional network protection is required.

Later, this same system can grow to a full-scale OC-48 system with full ring redundancy when revenues justify this additional network expansion. Within this migration path, SONET will require no redeployment of fiber. The OC-l2 system initially established and later migrated to an OC-48 network will utilize the same fiber without the need to lay any new cable. And since an OC-48 system can provide for the transmission of up to 96 channels of broadcast entertainment services, in most cases, an OC-48 will offer more capacity than what might be needed for traditional broadcast entertainment services. In the case of a 60-channel system, only two-thirds of the available OC-48 capacity will be utilized. The remaining capacity would then be available for other services.

MPEG-2 on ATM

SONET and ATM switching are well prepared to grow into the "information superhighway" and to provide the high-speed data and digital voice transmission of the future. It is appropriate to investigate whether the new digital video standards and especially MPEG-2 will fit into the SONET/ATM hierarchy.

In 1988, the CCITT formally defined the ATM cell formats (CCITT Standard I.361). As we can see from Fig. 3-4, the ATM cell format has a 48-byte payload and a 5-byte header.

The CCITT also defined the following layers of ATM:

*   The physical layer, which is concerned with putting bits on the wire and taking them off again.

*   The ATM layer, which handles cell mutiplexing and assorted housekeeping functions (such as header error correction).

*   The adaptation layers (AALs), complex sublayered protocols that package various kinds of higher-level user traffic into 48-byte ATM cells.

To achieve the ability of converting various data formats into the ATM cell format, the CCITT developed several classes of ATM adaptation layers - AAL 1, AAL 2, AAL 3/4 and AAL 5.

The MPEG-2 transport packet standard has been established at 188 bytes (often called octets) and, unfortunately, this standard was created with apparently little or no regard to the ATM standards.Thus, the question becomes: How to get the MPEG-2 transport packets into the ATM cell format - i.e., use which AAL layer?

The ATM Forum is now considering several methods that were presented in a January 1994 meeting.

If AAL-1 is used, four ATM cells will be required to carry one MPEG-2 packet. However, AAL-1 does not provide for a cyclic redundancy check or for a forward error-correction field. Another approach is to use AAL-5. The ATM cell payload for AAL-5 is 48 bytes and requires five ATM cells for an MPEG-2 transport packet. This is insufficient since 44 bytes in the fifth cell will remain unused. Two MPEG-2 packets could be adapted into eight ATM cells and provide forward error-correction and cyclic redundancy check.

So, as of the time of this writing, there is no standard for MPEG-2 compressed full-motion video to be carried through an ATM switch. However, there will be a standard. The ATM Forum, which consists of I5O companies, will certainly determine a standard before too many MPEG-2 boards will be shipped.

Reflections

SONET synchronous optical network architectures with ATM switching will form the backbone of the information superhighway and of any future long-distance voice, data and video traffic.

What are the implications of these new technologies to an enterprise?

*   If there are sufficient copper pairs, the internal voice traffic will remain in analog form and translate to digital for connection to the outside world.

*   In an upgrade program to fiber-optic, high-speed data, SONET with ATM switching within the campus, for campus interconnections and for long-distance is a recommended migration.

*   Video transmission requires a wide bandwidth. A SONET optical carrier OC-48 can transmit 96 channels. The expense of an OC-48 circuit may be justified for campus-interconnect links. SONET is, however, not an economical solution for video transmission within a campus.

*   The HFC hybrid, fiber/coaxial broadband architecture can carry video in analog or digital format. HFC can also carry voice and data within the enterprise and is by far the best buy for the money

*   MPEG-2 can be transported on the HFC system at any time in the future. The MPEG-2 transmission would be carried on coaxial and fiber-optic cables to the headend, where an ATM switch takes the MPEG-2 packet and transports them on a SONET circuit to the outside world.

* Until MPEG-2 television sets are available at today's TV set prices, all bi-directional video transmission will remain in the analog NTSC format. Any transmission to the outside world passes through the headend or gateway and is translated to T-I, DS-3 or MPEG-2 for long-distance travel.

* The gateway or headend of an enterprise offers the opportunity to combine voice, data and video into ATM cells for transmission on a SONET public network or to associated enterprise locations on a SONET private network.

Chapter 4

The Gateway and Operations Center

Whether the traffic consists of voice, data or video, every enterprise network must have a beginning. This strategic location consists of equipment complements to establish the functionality and operability of the network. Whether co-located with voice or data, or at a separate location in the enterprise, the operations center may be called the "Headend", the "Gateway", the "Video Operations Center" or the "Network Center".

Whatever the name, the operations center combines all the equipment necessary to establish the various operational functions to manage the network, to supervise and control all activities and to test the quality of the broadband transmission on a recurring basis.

To the cable television companies the begin location of their video delivery system is the "headend". The term stems from the equipment concentration required to receive and process TV channels. This equipment complement is responsible for all channel assignments and, therefore, the "headend" of the distribution system.

Since future voice, data and video communications will pass through the headend to interconnect with other common carriers or with other headend interconnect networks, the "headend" of the future will no longer be an end location but a bi-directional processing location.

If we transfer this terminology to the enterprise, it is appropriate to use the expression "headend" or "control center" for all intra-enterprise video traffic. For voice, data and video traffic leaving the enterprise, the headend becomes the "gateway" to the outside world.

The Control Center
for Intra-Enterprise Traffic

The Functions of the Control Center

The control center is the heart of the broadband HFC network and combines automation, origination and switching functions for both incoming and outbound voice, data and video transmissions.

For intra-enterprise data communication, the control center is the translation point of incoming data streams to outbound frequencies. Both voice and data translations follow IEEE 802 specifications and require a 192.25 MHz off-set frequency. Video channels do not have to follow this rule and are typically demodulated to baseband, switched at baseband and modulated to the RF channel of choice, in the outbound direction.

Fig. 4-I shows a typical video control center for two-way video transmission. Even though the equipment layout has been kept to a minimum, the functionality is quite impressive. Under computer control, permitting programmable event scheduling, the user can

    a)   select three programs for a cable company service feed

    b)   switch these programs to any of eight outbound channels

    c)   record two of the programs received on VCRs for automated retransmission on any outbound channel

    d)   program a video channel back to the cable company for redistribution in the cable network

    e)   originate information channels from a character generator

    f)   provide up to four outbound channels (Mod. 9-12) for simultaneous retrieval of video programs from four VCRs and one laser disk

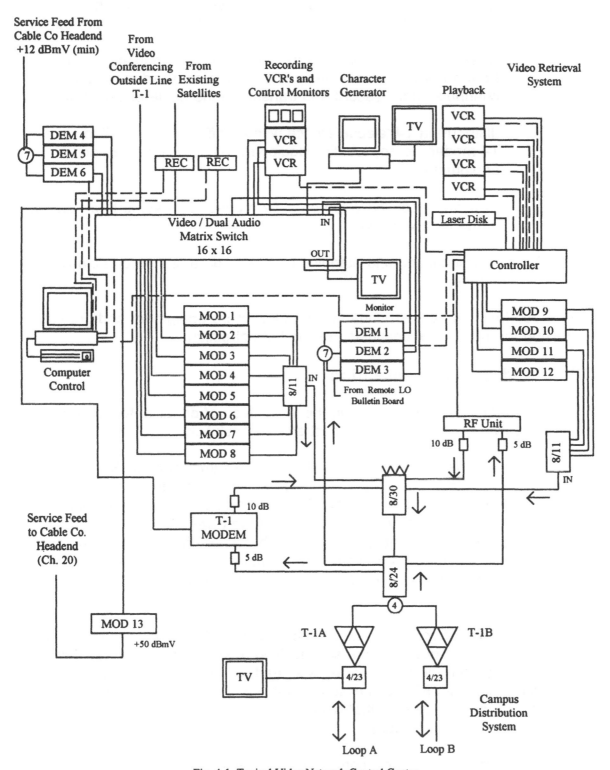

Fig. 4-1 Typical Video Network Control Center

from selected locations on campus

g)   receive programs from two satellite antennas and switch to
       selected outbound channels and/or record under computer control

h)   receive three simultaneous transmissions (demodulators I-3)
       from anywhere on campus and switch to selected outbound
       channels and/or record and transmit to the cable company feed

i)   distribute a T-I teleconference with an off-campus location via a
       T-I modem to any location equipped with T-I teleconferencing
       equipment

j)   view any incoming video on a TV monitor connected to an output
       of the matrix switch

k)   view incoming and outgoing programs on a standard cable-ready
       TV set

All of these functions are expandable by adding more equipment and gradually
increasing the number of channels used in both directions.

Off-air and Satellite Reception

If entertainment TV is a requirement, the installation of off-air antennas and
satellite receivers may be the most cost-effective solution.

Cable companies derive their channel plans in the same manner. But when
asked to provide these channels to your enterprise, the monthly fee for service
to your enterprise equals the bulk-rate that has been established for apartment
dwelling units. The license agreement between the town and the serving cable
company specifies apartment rates that were based on the costs that the cable
company incurs when wiring an apartment complex. Since you are spending
the money for the broadband HFC network implementation, the rates are
usually overstated.

Another deficiency of connecting a service feed from the cable company to your
network center is the quality of the signals. Your enterprise may be located 20
amplifiers away from the cable company's headend and the noise build-up is
approaching minimum FCC carrier-to-noise levels. So, unless a fiber-optic feed
is offered, you cannot be sure that what you will get is any better than the
house adjacent to your campus.

Another problem with the cable company's service feed is the channel lign-up.
Obviously, you must be in charge of your own spectrum management and,
therefore, would like to assign the channel space for the entertainment
channels on your system. Cable companies, however, like to give you the basic

84

and super-basic services the way they are used in the system, perhaps even with holes for eliminated premium channels. If you are told to do your own conversion of any channel that you would like to change the frequency assignment, you will find that between demodulators and modulators you will spend as much money as you will need to build your own receiving facility.

Of course, you may have a competing telephone company in the area that is just waiting to get permission from the Congress in Washington to provide cable services and all the above concerns may vanish overnight.

But if you are not in the position to wait, the prudent decision is to build your own receiving facility. There are a number of local satellite service providers that are agents for satellite programmers. Shop around for the lowest monthly rate for any channel that you want to receive. For instance, a CNN and Headline News can be priced as a package to obtain the lowest rate, per month and per active outlet. If you pick the 30 most important channels, the total monthly expense per outlet should not exceed $8.00.

Different programs are on different satellites. A total of three or four satellite antennas may be required to receive the channels that you desire to see on your system. The local satellite service company knows which program is on which transponder and on which satellite. Program suppliers change transponders and satellites often. Only your local satellite service companies will know the latest line-up.

Your equipment complement consists of off-air antennas, one for each local station with signal processors that convert the broadcast channel to the channel on your system.

Each satellite antenna has a low-noise receiver for both the horizontal and vertical polarization of the geostationary satellite. The next unit is the receiver, which can be tuned over the 500 MHz satellite band to pick out and demodulate one channel to baseband.

The receiver must contain a videocypher to decode any encoding protection that may have been imposed by the satellite service. Again, your local satellite service company knows which service needs to by deciphered and will arrange for your service agreement and associated paperwork.

The implementation of the off-air antennas, satellite dishes and downleads may be a separate turnkey contract with one of the local satellite service companies. Electronic equipment such as receivers and signal processors, which will be located in your control center, may be combined with other automation and switching equipment supplied under a separate contract.

While it is not the subject of this book to deal with the quality and longevity of off-air antennas and satellite dishes, it is common knowledge that satellite service providers are used to providing services to apartment complexes and small private cable systems. In this arena, the equipment selections are too often made on price and the overall quality of installation may not meet commercial standards.

Automation

It is most important that all operational functions be automated. Automation should not only cover real-time events, but also be capable of programming events over extended periods of time or up to three months in the future. PC-based software in the Windows format must permit the interlinking of numerous functions. For instance, a recording event first requires the selection of a channel by the demodulator. Once the demodulator is tuned to the desired channel and the signal arrives at the input of the matrix switch, the switch must be closed to feed the signal to the selected recording unit. The recording unit is turned on and the program is recorded for the prescribed time interval.

Automation software, in the same manner, should be capable of scheduling and directing any video retrieval and video-on-demand requirements. While it is of lesser importance to automate outbound frequencies of modulators, it is important to include all demodulators, VCRs, CD-ROMs, character generators, satellite receivers and the matrix-switch operation in the automation process.

Even with a high degree of automation, the operating technician still has to worry about the readiness of tape material, CD-ROM change-out and all other program origination functions. With the arrival of hard-drive digital storage devices, the future may offer relief from the task of inserting and removing VCR cassettes.

Recording of Programs

Whether from off-air, satellite channels or from any in-campus activity, the recording of video/audio programs is an important factor in the development of training material. The material can be edited, transferred to digital formats, archived, stored for delayed play-back or edited into sequences for interactive CD-ROM applications.

The recording process, when automated, does not require any human assistance. Educational programs of interest can be scheduled for recording during the night or on weekends without creating a staffing problem.

The volume of program recording events dictates the number of recording devices that must be available. While video storage is going through a transitional period, the most cost-effective recorder is still the VCR. Inexpensive VCRs count frame numbers, a somewhat inaccurate form of measurement, because of tape slippages in the start and stop processes. Different makes have different start-up mechanisms and can amplify the problem of playing a tape in two different players.

More expensive industrial machines are equipped with time-code equipment, which stripes the tape for location identification. Laser disks have accurate frame-number displays, because there is no slippage of a tape feed. Laser disk recorders, however, are expensive and usually only available at service centers that take your tapes and record them on a laser disk.

If there are programs that have a long-life cycle, it is recommended to record them on a multiprogram laser disk. Your automation equipment can access any desired frame number and play exactly the number of frames that are important for the lecture.

Storage of Video Programs

One-half inch video tape is still the most cost-effective storage media for video. When we, however, look to save storage space, tape is a very space consuming storage medium. Some installations feature archiving centers with eight-foot high rail-mounted storage cabinets.

Like paper storage, video tape storage will soon reach the end of its lifecycle. But what is there to replace it? With a requirement of 30 Megabytes of data for one second of video, we are talking about storing and retrieving massive amounts of data. Since post-production houses have a great need to get a handle on the storage of video programs, there are a number of new solutions approaching the market. For long-term backup or archival storage, there are magneto-optical drives, CD-ROMs, data cartridges and laser disks. For video-on-demand requirements where several users need access to the same program, installing a video server makes a lot of sense.

The compression of video is a related issue and needs to be included into the video-storage equation. An application of MPEG-2 compression standards can reduce the data requirement for one second of video to 1 to 2 Megabytes. MPEG-2's more sophisticated compression method allows for all the data in a full CCIR-601 picture frame to be compressed. While removable media offers greater archival storage capacity, it is not fast enough to handle the high data transfer rates. It is possible that fixed hard-drives are more suitable for MPEG-2 encoding because of their higher transfer speed.

Considering the massive effort that is being expanded by the industry, MPEG-2 equipment will be available soon at reasonable prices. At that time, CD-ROM writers with 650 MB of storage, and optical libraries with up to 170 GB of storage capacity, may finally replace the video tape.

Whatever your video storage needs may be, it may be prudent to wait a year or two before investing substantial funding into digital video storage. The application of digital storage, however, does not mean that your HFC network must carry video in the digital format. A simple conversion to analog NTSC will permit viewing of the video on standard, inexpensive TV sets and an NTSC to VGA conversion card in the computer will add to the versatility of the system. It is a remarkable example of the current state of solid-state technology that such complicated operational capabilties can be fabricated on a single chip. Both digital-to-analog and analog-to-digital conversions will soon be common place.

Channel Assignment

In order to be able to manage the spectrum of your broadband HFC facility, it is important to be in charge of the assignment of any inbound and/or outbound transmission. Assigning a channel means determining the RF frequency of the transmission. In order to be able to assign channels, the video is switched to the output port of the matrix switch that feeds the modulator, which is preset to the desired RF channel.

The matrix switch, therefore, must have as many outputs as there are assignable channels. This can lead to large matrix-switch configurations and costs. The compromise that is usually made is to preselect the channel plan of entertainment TV-channels and let them bypass the matrix switch. This will reserve the crosspoints for local origination, two-way video transmissions, videoconferencing and transmissions to and from the outside world.

Editing and Authoring

Whether you are a high school, university, hospital, a large manufacturing facility, the training requirements are increasing exponentially. And since pictures, and especially moving pictures, are worth many words, an interactive video/audio learning platform is essential in the future.

This does not mean that your enterprise has to create large video postproduction linear editing facilities. It means that some form of video editing must be available for program dessimation of educational videos and for integration in interactive, multimedia authoring stations.

Nonlinear editing stations have matured during the last years and are surprisingly easy to handle. Nonlinear editing of video sequences and integration of such sequences into a multimedia, interactive authoring program promises to become an efficient learning environment.

Again, the CD-ROM production that you just put together contains NTSC full-motion video sections and cannot be downloaded over an ethernet LAN. A digital-to-analog conversion, however, permits switching to an RF-channel and transportation to the selected workstation over the HFC broadband network and without any quality degradation.

This is not to say that every educational video program will be coming from authoring stations. Video/audio transmissions to TV sets that are equipped with control elements that let you stop, play, fast-forward and pause a program will become a formidable aid in the educational process.

Program Origination and Scheduled Programming

The control center of the HFC network is responsible for all program originations. There may be a number of information channels that inform about programs, program content, channel number and time of start and duration information. These programs may consist of live or stored programs of educational value to the enterprise staff. In educational facilities, these programs may consist of replays or recorded lectures. In a hospital network, multichannel programming may be provided for the training of paramedical personnel, nurses and even physicians.

The play-back equipment, again, may consist of old-fashioned VCRs and may be converted to digital storage and CD-ROM play-back in the future. Computer control has the ability to time-program all transmissions so that the play-back starts on schedule and is switched to the selected channel that has been announced on the information channels.

Video Retrieval and Video-on-Demand

The term video retrieval signifies the ability to turn on and control a video program from selected locations on the HFC network. The term video-on-demand (VOD) was coined at some point later in time when the cable companies discovered that the user may also want to select a video rather then having to chose from a preselected group.

Video retrieval then depicts a preselected group of videos that have been inserted into VCRs and are ready to go. The lecture hall, classroom or conference room is equipped with a retrieval control unit, which permits the selection of one of the programs and allows the presenter to manipulate the video and make it an integral part of the lesson plan.

Video retrieval communication requires a two-way signal stream on the HFC network, usually working at around 5 MHz in the inbound and around 220 MHz in the outbound direction. The communication with the control unit at the control center can also be established over telephone lines and DTMF transmitter/receiver units in the control unit. This method may be used from resident student dormitories not connected to the two-way HFC system and receiving only entertainment services.

Video-on-demand adds the dimension of selection to the process, which may require a substantial number of play-back machines ready to go. Video tape players cannot handle this kind of demand. Also, multiple requests for the same program cannot be satisfied using tape players. Video-on-demand services, however, can always be added to the HFC service menu at a time when digital storage is available at reasonable prices.

The Interactive Classroom

Video retrieval and the ability to control a video program from a handheld remote control unit permits a userfriendly transition from a presentation classroom to an interactive classroom. While the standard presentation lecture hall relies on local production equipment such as VCRs, cameras, workstations with NTSC outputs or CD-ROM players with NTSC boards, the interactive classroom can also retrieve video material from video storage at the control center.

In addition, the interactive classroom can conduct teleconferences with any other lecture hall or even with distant locations.

Fig. 4-2 shows a computer lab that is connected to the HFC system and incorporates video presentation equipment and video retrieval.
The instructor can

a) retrieve and control any video sequences from the material at the control center

b) originate local video from camera, VCR, workstation or CD-ROM

c) receive video transmissions from other or distant locations through the broadband cable connection

d) transfer any programs of (a), (b) or (c) to the projection TV sets and to any connected computer workstation. All other workstations are interconnected by RGB daisy-chained wiring

e) every computer workstation, under the control of the instructor, can also transmit to the projection TVs or to a distant location via the HFC network

f) monitor the progress of each student by viewing the screen content, switch his screen to all students, project a student's screen on a TV set or transmit it to other participating locations.

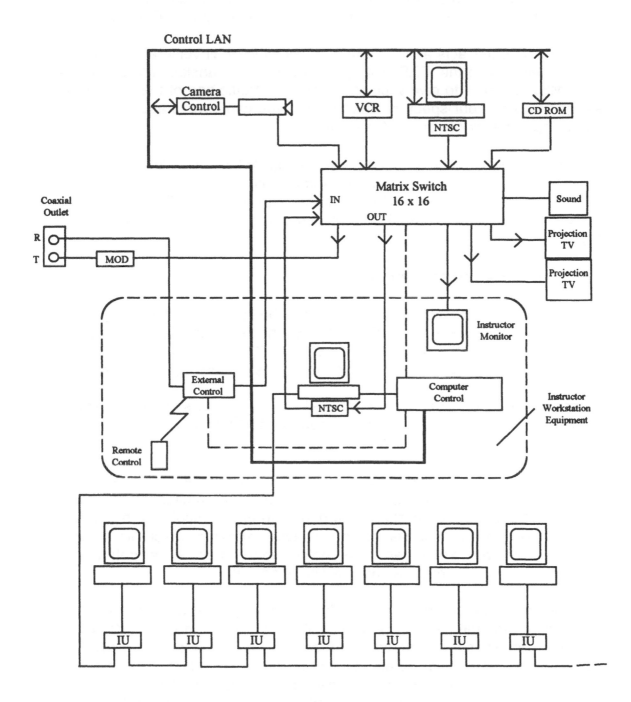

Fig. 4-2 Interactive Multimedia Computer Lab

Voice and Data Traffic

There are T-I cable modems available that will take 24-voice channels or I.544 Mbit/s data to an RF-channel assignment for travel on the HFC system.

The control center acts as the turnaround location of all voice and data traffic. In accordance with the IEEE 804 specifications, the translation frequency is I92.5 MHz. Be sure to give your voice and data traffic one of the frequency assignments on your spectrum chart so that you can always assess the remaining capacity of your HFC system.

Fig. 4-3 shows equipment required and the signal flow of intra-enterprise and gateway handling of voice and data communications on the HFC system. The T-I format is especially well suited for transmission on the HFC broadband since it is easily integrated into an ATM-switched SONET ring for the long-distance segment.

Desktop Videoconferencing

Full-motion video at NTSC specifications can be incorporated into any computer terminal by fairly inexpensive VGA-to-NTSC and NTSC-to-VGA conversion cards. This means that within the area of an HFC broadband system all videoconferencing can be provided to the highest picture quality standards.

Present workstation video follows MPEG-I, or even lower standards, permitting picture frame rates of I5 frames per second and resulting in a jerky presentation.

Digital video presents even more difficulties for LANs because it has characteristics that are foreign to most. First, LANs are designed to handle data in a democratic manner, assigning no priority to any particular type. Second, data typically travels in bursts, not in continuous streams, which are necessary for video. Microprocessors tend to be optimized to handle either data bursts or data streams. Few, if any, can handle both efficiently.

It is, therefore, not surprising that desktop videoconferencing has not seen a lot of growth during the past years. The availability of a broadband HFC network, however, simplifies the requirement for continuous transmission greatly. Participating workstations screen content is converted to NTSC video and modulated to an RF frequency for transport to a routing switch in the control center. The return path to other workstations is selected in the routing switch and NTSC video is reconverted to the VGA format of participating workstations.

The quality of the video is of great importance to telemedical and diagnostic video presentations. The medical community has rejected video transmissions at speeds lower than T-I to be of any diagnostic value. T-I transmission, of course, can be established as well on the HFC network except that the present day price tag is still high. The development of desktop videoconferencing within an enterprise is seen as an analog NTSC high-quality transmission because standard TV sets can also be used as inexpensive add-ons for workstations.

Network Management

The supervision of the HFC network performance is a basic requirement. All traffic, whether video, data or voice, is accessible at the control center. Here are some of the monitoring tasks that should be checked, as a minimum, in a periodic manner:

a) The inbound and outbound picture quality

b) The fiber-optic transmitter and receiver power levels at the control center and at fiber-node locations

c) The incoming RF levels from on-campus origination locations or desktop video locations

d) The T-I telephone and data traffic

e) The record of outbound video transmissions

f) Recording events and video retrieval requests

g) The number of simultaneous return transmissions

h) Items (f), (g) and (h) will provide useful information relative to system usage and future needs

The key to a comprehensive data collection is a software that has been customized to your particular needs. Since all scheduling events are already recorded to memory as part of your automation software, the network management software must integrate the information of test equipment, analog sensors and digital counters into the network performance report document.

The Gateway to the Outside World

The Functions of the Gateway

While Fig. 4-I dealt mainly with intra-campus video traffic, Fig. 4-3 puts more emphasis on connectivity with the outside world. Fig. 4-3 also shows interconnections of the control center with an HFC fiber-star network to and from fiber-node locations.

RF modulators and T-I modulators are used to convert video and audio baseband to RF channels in the outbound direction of the broadband HFC network. The video baseband either comes from output ports of the matrix switch or from T-I lines from the outside world. In our example, one of the T-I lines is a distance learning connection to a distant facility. The other T-I video comes from an ATM switch that is connected to a SONET ring of the information superhighway.

Other ATM connections in T-I or MPEG-2 format are decoded and the baseband connected to input ports on the matrix switch for distribution over the campus HFC system either by RF modems or by T-I modems.

RF channels received at the four fiber-optic receivers are combined in one RF spectrum and then distributed to T-I demodulators for baseband transmission to the ATM switch. Other NTSC video RF channels are demodulated to baseband and appear as an input on the matrix switch.

The output of the matrix switch feeds RF or T-I modulators for retransmission throughout the HFC network. Other outputs are provided to feed the baseband video to T-I and MPEG-2 codec equipment, which is interconnected with the ATM switch for transmission on the SONET long-distance network.

The voice and data translation equipment changes the incoming RF carrier to an outbound carrier I92.5 MHz higher and sends any T-I voice or data back through the HFC system on a discrete channel. Intra-campus voice/data and even videoconferencing traffic have their own RF channel assignments, which are different from intercampus and long-distance channel assignments.

Distributed Distance Learning

The providers of distance learning and multipoint teleconferencing equipment always ask where the distance learning classroom is and then proceed to install copper or fiber wiring directly to the location. If you wanted to move the equipment, it could not be done without expensive moves and changes.

The reason for this inflexibility is the desire to sell you a rack of equipment, which includes the codec. Once you have your HFC network in place, you can begin to think in terms of a distributed architecture.

Fig. 4-3 shows a regional T-I outside line and an ATM-derived T-I circuit. All that needs to be done is to modulate the T-I to an RF channel on your HFC system and you can move your distance learning equipment to any outlet anywhere in the system using a T-I modem in front of the distance learning equipment input port.

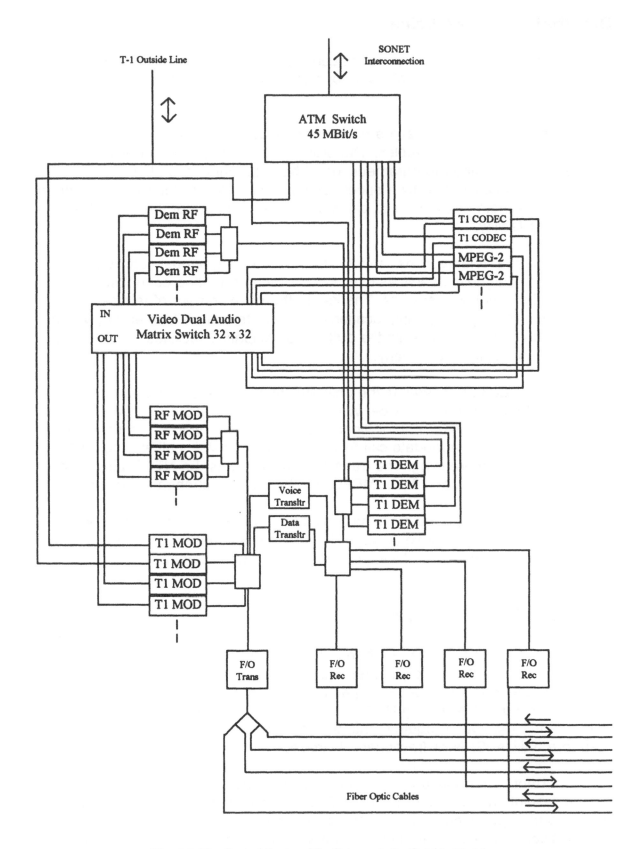

Fig. 4-3 The Control Center - The Gateway to the Outside World

Distributed Telemedicine

In an effort to bring the cost of health care delivery in line with only inflationary increases, telemedicine is praised as one of the cost-cutting measures of managed care.

Many hospital complexes have a number of associated medical facilities within a few miles of the main complex. The provision of health care in these outlying facilities is often limited to paramedic or, at best, primary care physician services. Any examination by a specialist would have to be deferred to a visit to the hospital's main complex or to particular visitation hours of the specialist at the outlying facility.

It is obvious that telemedicine, and the ability to instantaneously consult the specialist for diagnostic services over a voice/data and video connection, would provide substantial cost savings. The key to the development of telemedicine is the selectability of participating locations. Specialists can be located anywhere in a hospital campus. So, while the outlying participating facilities are few, the intra-campus locations are many.

The HFC broadband network is the only architecture that can comply with this requirement for multipoint service. What is required is a video exchange similar to the familiar voice PBX, but for video, data, sound and voice. The software must be able to accommodate call forwarding and transfer, call waiting, call screening and conferencing. The transmission path must be NTSC within the hospital campus area and T-I or MPEG-2 in the local exchange segment.

There are few video phone systems available at the time of this writing, which feature interesting control software and operating parameters. By the time you have your HFC network in place, you should have no problem securing a cost-effective solution.

Desktop Videoconferencing

It is interesting that the video exchange for the multipoint requirements of telemedicine also fits the requirements of a desktop videoconferencing network. While the interconnection between participating workstations within the campus is best handled in the analog NTSC format, connections to the outside world do not have the resolution requirements of telemedical connections and can, therefore, utilize ISDN or fractional T-I circuits.

It is important, when evaluating desktop video phone systems to assure that picture resolution is in the range of 720x480 pixels, that an auxiliary NTSC port (in and out) is provided, that 30 frames per second are used in the transmission, that a minimum of 4 to a maximum of 16 users can be seen on a screen and that both T-I (V.35) and ISDN interfaces exist. In addition, the equipment integration at the desktop should include an integrated camera, speaker, microphone, speaker phone and headset, and an audio mixer for combining multiple audio channels.

Telephony and Data

The broadband HFC network can carry T-I transmissions anywhere throughout the campus area. Within a building, the installation of copper for additional telephone requirements is the most cost-effective solution. But a T-I multiplexer and a T-I RF modem at building-entry point saves the expense of copper or fiber cables. The 24-voice channels are simply carried on an RF channel on the HFC network to the control center. There they are translated to an outbound channel assignment for transmission to the PBX.

The same approach can be taken from the existing PBX switch to the ATM switch at the control center to reinforce the toll capabilities of the system.

By associating fractional T-I segments to certain locations, it is, of course, possible to develop a muiltipoint architecture for both additional voice and data requirements. While inside a single building, the utilization of separate facilities for voice, data and video is the more cost-effective solution; within a campus area, the HFC architecture permits economical transmission of anyone or all of these three in analog or digital form.

Personal Communication Service (PCS)

A broadband HFC network is the ideal conduit for a PCN - personal communication network.
A PCN is the network that is required to get personal communication services from the switch to the user and back. A PCN employs small coverage cells of less than a 1000 ft. area. The cell transmitter has output powers of between 0.01 to I Watt. The handheld unit has a transmit power of only 10 mW.

Herein lies the main advantage of the PCS service. Low-power PCS phones will be able to fit into your shirt pocket. On the other hand, many more cell transmit and receive locations are required to provide uninterrupted coverage at any location within the system.

To interconnect the many transmitter/receivers with twisted pair would require a new maze of copper wires, which is not a feasible approach. But interconnection of these cells by RF on a broadband HFC system makes imminent sense. The RF modem connects every one of these cells with the gateway control center where a PBX is located to handle the enterprise's internal telephony. Connections with the outside world are handled as usual, i.e. by interconnection with the local exchange carrier via T-I multiplexers or even through the ATM switch.

Chapter 5

The HFC Broadband Network Components and Performance

HFC Network Architectures

An HFC network always utilizes single-mode, fiber-optic cables for the transport segment of the network and coaxial cables for the signal distribution in the last mile. The reasons for this architecture are discussed in the following.

While the coaxial distribution segment is always a tree-and-branch architecture, the fiber transport section can be arranged in either a star or ring configuration.

The Fiber Star and Coaxial Tree-and-Branch

Fig. 5-I shows the principles of the fiber star transport system and the coaxial tree-and-branch distribution system.

The equipment complement at the control center or gateway has been oversimplified for this presentation. The HFC control center is the routing, conversion and switching location for all voice/data and video traffic entering or leaving the HFC area.

The control center is the central office of the broadband network. Voice/data and video transmissions from within the HFC area, and intended for participants within the HFC network area, will be received and converted to outbound frequencies. Voice/data and video transmissions received from within the HFC area, with long-distance prefixes, will be converted to digital transmission for transmission through ATM switches and SONET interconnections. Incoming long-distance transmissions will be converted to the format required (analog or digital) by the subscriber, modulated to the assigned RF frequency and transmitted in the outbound direction.

The control center or gateway of your enterprise performs identical functions, just on a smaller scale.

The transport segment of the HFC network consists of optical transmission. At the control center, there is a concentration of high-power, fiber-optic transmitters required to illuminate all outbound fiber strands leaving in many directions. Optical splitters at the output of the fiber-optic transmitters are used to feed multiple fibers in the most economical configuration.

The transport segment ends at the fiber-node location where the optical bandwidth is translated into electrical RF bandwidth. The distribution segment consists of coaxial cable with amplifiers, passives and multitaps for service to the subscriber. The subscriber receives the entire RF bandwidth in the forward direction. Special equipment at the subscriber premises is required to view television, receive a data stream or talk on the telephone.

In the return direction, a cable modem is used to locate the subscriber's voice and data transmission in an RF frequency range suitable for return. In a sub-split system, this frequency range would be 5 to 30 MHz. Since this narrow bandwidth is not capable to transport many video channels simultaneously, there are voices that suggest that any future video requirement from the home may be transmitted at 600 MHz and higher.

A high-split architecture, as the one recommended for your enterprise, could be built, but the cable companies feel that it is important to transport Ch 2 on channel, i.e on 54 to 60 MHz. Since most every cable operator uses "channel mapping" in their set-top converters, this reliance on using the forward band from 50MHz in the outbound direction may not be justified. Channel mapping is a simple addressing concept, which allows a Ch 2 set-top converter indicator to be set for any incoming signal frequency. So, the Ch 2 designation could well be used at 222 MHz in a high-split system.

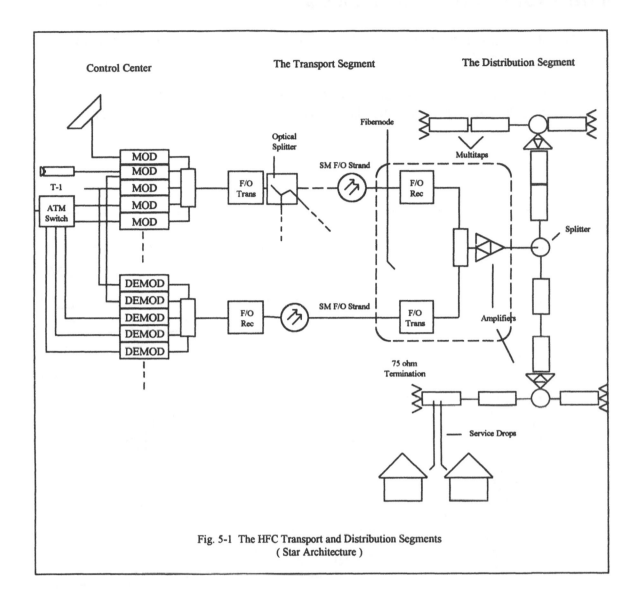

Control Center **The Transport Segment** **The Distribution Segment**

Fig. 5-1 The HFC Transport and Distribution Segments
(Star Architecture)

Whatever the return spectrum of the future broadband HFC system may be, the entire band can be transmitted over the same coaxial cable to the fiber-node location. This means that only one single drop wire is required for forward and return transmissions.

This single-wire architecture stops at the fiber-node where a fiber-optic, low-power transmitter is required to illuminate a separate fiber strand for the optical return to the fiber-optic receiver in the control center. Again, the optical transmission is converted to electrical transmission and demodulated to its baseband format.

The Fiber Ring and Coaxial Tree-and-Branch

Fig. 5-2 shows an example of an HFC system utilizing a ring architecture in the fiber-optic transport segment.

It is noted that control center functions and distribution segments are identical to the fiber-star system. The only difference is the routing of the fiber-optic cables.

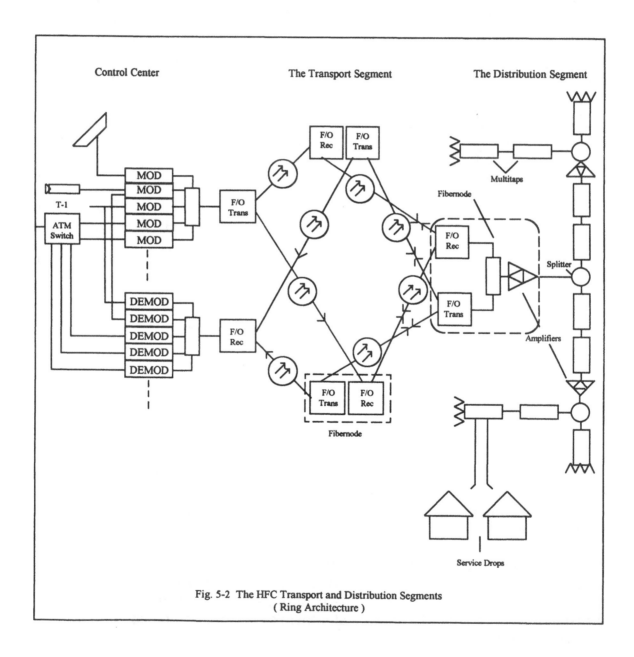

Fig. 5-2 The HFC Transport and Distribution Segments
(Ring Architecture)

The ring architecture provides for route diversity and, therefore, higher circuit reliability. Every fiber-node can receive the signal from two directions and can transmit the return signals in two directions. This may be an important consideration in aerial outside-plant areas where outages, due to weather and vehicular traffic, can be frequent.

Having more routes to take means more fiber-optic cables and, therefore, a somewhat higher price tag. But when the circuit is not just a voice pair, but a broad spectrum of services to a group of subscribers, the additional expense may well be justified.

The Similarities of Fiber-optic and Coaxial Cables

The Spectral Capacity

Both single-mode, fiber-optic strands and coaxial cables have equivalent spectral capacities. The emphasis is on single-mode fiber operating at 1310 nm. Multimode fibers cannot transmit many video channels at one time and have a considerably higher attenuation.

The spectral capacity of both coaxial cable and a single-mode fiber strand is in the range of 1 GHz. The equipment that is being used today is not yet able to meet the capacity that is offered by the two media.

In the case of the single-mode fiber cable, transmitters for 50 to 750 MHz are just coming to market. Amplifiers for coaxial cables are available for the same frequency range but still in design for frequencies over 750 MHz.

 A spectral capability of 700 MHz or 116 6 MHz wide analog TV channels with stereo/audio is overwhelming. Compare this spectral power with an LAN by translating bandwidth to speed - using only a requirement of 45 Mbit/s per video channel - the required bit-stream speed would be 5.2 Gbit/s. Using MPEG-2 compression in this example, and assuming that the standard will settle at 3.0 Mbit/s, the 116 channels still would require 348 Mbit/s. Still a little bit out of reach for high-speed data systems and also too expensive. Since it is state-of-the-art to locate six T-1 channels in a 6 MHz video channel assignment, this also means that both a one-directional fiber and coaxial cable could carry 696 T-1 circuits or (696x24) 16,704 voice circuits.

But the more convincing argument is that voice/data and video circuits can live side-by-side on both fiber-optic and coaxial cables. All they require at interface points is a photo diode to translate from light to electrical and a laser modulator to proceed from electrical to optical transmission.

Amplitude RF Modulation

Telephone companies increase bit rates of digital transmission systems to carry more traffic. An OC-I can carry 672 voice channels. A OC-48 can carry 32.256 voice-grade conversations. When it comes to broadband transmission, the cable companies have always increased the bandwidth.

This old technology for assigning a frequency for a broadcast transmitter goes back to AM radio. An RF carrier is produced at the desired transmitter frequency and the amplitude modulation of the signal is imposed on the carrier. That is what broadband is and that is the principle of HFC technology.

When the FCC determined in the 1930's that a television broadcast carrier must be single side-band and fit into a 6 MHz frequency band, they invented broadband cable TV systems and the HFC system of the future.

In chapter 2, we explained in detail what this broadband spectrum is capable of. Both fiber-optic cables and coaxial cables can handle such a broadband spectrum, with RF carriers every 6 MHz, well and without any difficulty.

Before translating the electrical signals to an optical transmission at 1310 nm, it is important that all RF carriers are stacked one above the other and that the broadband output has been combined into a single cable. This cable becomes the input to the fiber-optic transmitter, which transmits the entire amplitude-modulated broadband at the optical frequency of 1310 nm. The photo diode in the receiver translates the optical broadband back to an electrical broadband for transmission on the coaxial cable.

Analog and Digital Transmission

At baseband, an analog transmission has various levels. A digital transmission, however, only consists of ones and zeros. This substantial difference between analog and digital transmission parameters can be overcome by modulating the RF carrier using amplitude modulation.

Neither fiber nor coaxial media cares whether the information on the RF carrier is in an analog or in a digital format. This offers the flexibilities that are needed for the development of broadband services.

- Cable TV services can be transmitted in analog form until inexpensive digital TV sets are available and use standard set-top converters

- Video-on-demand services can be transmitted in the digital MPEG-2 format and made available to the subscriber via a special D to A set-top converter

- Telephony, data and even high-speed data can be transmitted in T-I or E-I sub-groups and in digital form via a cable modem, which can interface directly with ATM switches and SONET long-distance networks

The HFC broadband network within your enterprise is not different from the HFC network now proposed by telephone and cable companies. Anything they can do, you can do. No FDDI or fast ethernet network can offer the flexibility of transmission that the HFC network can provide.

The Differences between Fiber-optic and Coaxial Cables

The Attenuation Difference

Single-mode, fiber-optic cable does not lose much illumination through dispersion at 1310 nm. Therefore, an optical transmission has attenuations as low as 0.35 dB per Kilometer. And since the entire 750 MHz wide RF band is traveling at the wavelength of 1310 nm, this attenuation is the same for all frequencies.

Coaxial cables have varying attenuations over the frequency range. The higher the frequency, the higher the attenuation. At 750 MHz a coaxial cable with a diameter of 0.9 inches has an attenuation of 1.29 dB per 100 ft. At 3.3 ft. to the meter, this would compare to 42.5 dB per Kilometer.
The choice is clear. Fiber-optic cable is the only medium that can overcome distances.

So, why use coaxial cables at all?

The Distribution Difference

In order to serve a number of users, the fiber-optic cable requires to split the optical transmission to transfer to other strands. The device is called a fiber-optic coupler. Coupling losses of 3.5 dB are typical for a 2-way split. Losses of 7.0 dB are incurred in a 4-way coupler.

If you consider that an optical link (transmitter to receiver) has a link budget of 10 dB, then you have a choice. You can span a distance of 28 Kilometers or you can split the fiber transmission 8 times and you cannot even go around the corner. The coaxial plant, on the other hand, has link budgets of 22 dB between amplifiers and can install 8-way multitaps with insertion losses of between 0.9 and 3.7 dB depending on the level that is required to be delivered to a subscriber.

An HFC system uses this basic difference between fiber and coaxial to its fullest advantage. Fiber-optic strands with optical transmission are used for the transport segment and coaxial cables with electrical transmission are used for the distribution to the many users.

The Power-Carrying Capacity Difference

Fiber-optic strands do not have a metallic conductor to carry any voltage for the powering of equipment. A IIO Vac supply must be provided for every equipment location.

Coaxial cables, however, have good power-carrying capacities because of the large gauge center conductor and the solid outer sheath. Cable TV amplifiers have been powered through coaxial cables for the last three decades. A 6O Vac I5 Amp. power supply can serve up to IO amplifiers in one power-supply area.

Service drops emanating from multitaps usually do not carry power to the subscriber's home and requiring the subscriber to supply power to set-top converters and other equipment. The telephone companies, however, have always supplied the ringing current for telephone sets from the central office, which raises the question about what to do in the broadband era of the future. Some cable manufacturers are already combining coaxial drop cables with one twisted pair to fill the void.

Optimizing the HFC Network

It is obvious that the installation of a single fiber-optic cable to each subscriber would be an expensive proposition. This method also is not justified from a usage point of view. Since the spectral capacity is IOOO MHz, less than I% of this capacity would be used by a single user even when he is receiving two video channels at the same time.

In the reverse direction, it is hard to justify even one video channel or a T-I connection. But even if we would assume that this is a possibility by the year 2O5O, the capacity used would be less than O.7%.

If each user uses less than I% of the spectral capacity, then IOO subscribers would utilize the full capacity of the system. But how many minutes per day would a subscriber use voice/data services? Video services are constantly supplied on the cable and only need to be tuned in, but voice and data, or even high-speed internet interfaces, require a frequency slot on the cable.

It turns out, even when assuming a MPEG-2 video transmission coming from every subscriber eight hours a day, only about 2O MHz of spectrum would be required. Based on this exercise, some future broadband service suppliers are saying that each fiber-node can serve up to 500 subscribers. Others are saying that the costs of serving less than 150 subscribers from one fiber-node cannot be justified.

What are then the important considerations in the optimization of an HFC network?

Whether we are talking about the new broadband cable TV company, the new broadband telephone company or your enterprisewide HFC system, the major considerations are

a) reliability
b) quality of service

The cable TV companies, in the early years of cable TV, built systems with 30 plus amplifiers in cascade. As a result, the reliability decreased because of amplifier, cable and power outages. Caused by the large number of amplifiers in cascade, the carrier-to-noise (C/N) ratio decreased and the composite triple-beat (CTB) of all the RF carriers increased to the point that the picture quality became substandard.

Cable TV companies earned a bad reputation for the frequency of outages and the noise or herringbones on the TV sets, which the industry to date has not fully overcome.

Obviously, these mistakes will not reoccur when using the HFC architecture.

Optimizing the Reliability

A good reliable circuit should have an availability of 99.999%. There are 8.760 hours in a year. 99.999% just about amounts to a total outage time of 0.1 hour/year or 6 min./year

It is interesting to note that equipment manufacturers have spent substantial resources on improving to MTBF meantime between failures of amplifiers and passive equipment. It turns out that when you use three amplifiers in cascade, you probably do not see one of them fail in a 5-year period.

It was the coaxial cable routing that caused most of the outages. Trunklines have the tendency to follow major arteries. One car taking out one pole can turn off the entire system for hours.

HFC overcomes this problem by sending a fiber trunk to each group of 150 to 500 subscribers. There can never be a total outage. The fiber route is still subject to weather and vehicular accidents, but it could be installed following diverse routes to offer redundancy.

The Reliability of the Enterprise HFC Network

The resulting availability of your enterprise HFC system may even be 99.9999% when considering that outside plant is underground and most of the coaxial equipment designed for outdoor use is installed inside of buildings.

The above percentage time of availability means that every fiber-node area with a total of three amplifiers in cascade will only have less than 0.6 minutes of outage time per year. The high reliability can be translated to cost-savings on maintenance. No routine maintenance of the infrastructure is required. Maintenance activities can be restricted to the various equipment connected to the HFC network.

In case of the unlikely event of a fiber-transmitter failure or of a coaxial forward or return amplifier, modular replacement can keep the (MTBR) meantime to repair within an hour or two.

The high reliability of an HFC system, however, depends on the integrity of the IIO Vac power service. It is recommended that secure and uninterruptable power service is used for the powering of fiber-optic transmitters. At fiber-node locations, all equipment such as fiber-optic receivers and the power supply for coaxial broadband amplifiers should be connected to a secure uninterruptable power source.

Optimizing the Quality of Performance

Looking at Fig. 5-I and 5-2 again, we can see that an HFC network consists of the transport and the distribution segment for numerous fiber-node areas. What is interesting is that the number of active components in each fiber-node area is the same.

Calculating the quality of performance data once, for one fiber-node suffices, as all other service areas will have almost identical equipment complements. This is valid, provided we have determined all the ground rules for our system.

Outlet Levels

Even though, the quality of service is not affected by the determination of the HFC outlet level, it is one of the most important specification. The outlet level describes the robustness of the system.

In the early days of cable television the FCC determined that the minimum level shall be O dBmV at a TV outlet. Our HFC system has to provide sufficient level for multiple services for transmit and receive as well as for all frequencies. The

HFC outlet level, therefore, should be in a range to allow for attenuation difference at the various frequencies and strong enough to permit further signal splitting. The recommended outlet level for a sub-split system is +10.0 ±4.0 dBmV and for a high-split system, which requires two-way characteristics, the outlet-level range should be tighter and in the range of +8.0 ±3.0 dBmV.

The Carrier-to-Noise Ratio

The carrier-to-noise ratio (C/N) defines the number of dBs that the carrier is located over the noise floor of the equipment. Nonlinearities in the electronic circuitry cause each equipment to have a particular noise figure. It is the effect of the noise figure that establishes the C/N of a unit.

In the early days of cable TV, a C/N of 39 dB was the minimum specified by the FCC. Experience, however, showed that visible degradation of the picture details can be observed if the C/N value falls below 43 dB.

In an HFC system, there are a number of contributors that affect the overall CNR of the circuit. These are

-       the fiber transmitter C/N
-       the fiber cable C/N
-       the receiver shot noise C/N
-       the receiver thermal C/N
-       the trunk amplifier C/N
-       the C/N of the distribution amplifiers

The FO Transmitter C/N

The C/N of a fiber-optic transmitter is determined by the noise that is being generated by the laser. C/N values vary greatly with the channel loading. Specifications are often based on a 60-channel loading and are in the area of 51 dB, considering a constant triple-beat performance.

The Fiber Cable C/N

The fiber-cable noise is directly related to the distance the light has traveled through the single-mode fiber, the optical power and a factor that is contributed by the type of laser in the transmitter. While the laser factor and optical power must be specified by the manufacturer, the distance factor can be added on a logarithmic basis. The formula for a distributed feedback laser is $10 \log (2/D)$ and for a Fabre Perot laser $23 \log (2/D)$ where D is the distance in kilometers (km).

The C/N of the fiber cable is then the addition of

Laser factor + distance factor + power factor

A typical C/N is a calculation for a 26 km distance and a power factor of 27 dBμW/ch minus a reference power of 17 dBμW/ch using a Fabre Perot laser as followed:

$$\text{Fiber C/N} = 58 + 23 \log (2/D) + 2x (27-17)$$
$$= 58 - 25.7 + 20 = 52.3 \text{ dB}$$

The FO Receiver C/N

The C/N of a receiver is composed of two components, the thermal C/N of the receiver and the receiver shot noise C/N. Chapter 10 discusses link budgets and receiver power levels in more detail. The receiver level is determined by taking the optical receiver power from the manufacturer's specification and using the formula:

$$\text{Receive level} = 2x (ORP-14) -1$$

Assuming that the optical receive power is 13 dBμW/ch, the receive power level will be 2x (-1) -1 or -1 dBmV. This means that the thermal C/N is 1 dB lower than the carrier-to-noise ratio on a typical receiver spec. sheet. Depending on channel loading, receiver C/N ratios are around 52 dB, which gives us a thermal C/N of 51 dB.

The shot noise level is, again, a function of the receiver location. The manufacturer's specifications should state a negative number in the high 40 dBmV range. Assuming this number is -47.2 dBmV, then, by adding the optical path loss to it, we obtain the receiver shot noise levels:

$$-47.2 \quad -13.0 = 60.2 \text{ dBmV}$$

To obtain the receiver shot C/N, we still have to apply the above receive level formula:

$$\text{Receive Level} = 2x (ORP-14) -1$$
$$= 2x (-1) -1$$
$$= -1 \text{ dBmV}$$

This deduction gives us a shot C/N of 59.2 dB.

Broadband Amplifier C/N

Coaxial amplifiers have been developed for many years with the goal in mind to increase the C/N ratio. The basic inherent noise figure of an amplifier has been reduced to 4 or 6 dB. Noise figures at the high end of the transmission band are still around 8 or 9 dB. The resulting C/N ratio is, therefore, a function of loading, frequency and the passband that is being transmitted. The worst case would be a sub-low forward system with an 80-channel loading operating between 50 and 750 MHz. The C/N will be in the range of 55 to 58 dB in any

4 MHz band. A high-split amplifier with less loading between 222 and 750 MHz has slightly better C/N ratios, i.e. between 57 and 60 dB.

The System C/N

Now that we have worked out each of the C/N ratios from the fiber-optic transmitter to the last broadband amplifier, we must determine the C/N ratio of the entire system.

The cable TV companies have a golden rule. Double the number of amplifiers and you will reduce the C/N ratio by 3 dB. So, when you have 16 amplifiers in cascade, the C/N ratio of the system reaches the subjective limit of 43 dB where impairment becomes noticeable.
But, in our case, we do not have 16 electronic devices and the C/N ratio must be far better. We have

> the Fiber Transmitter C/N = 51 dB
> the Fiber Cable C/N = 52.3 dB
> the Receiver Thermal C/N = 51 dB
> the Receiver Shot Noise C/N = 59.2 dB
> the Trunk Amplifier = 55.0 dB
> the Distribution Amplifier No.1 = 52.0 dB
> the Distribution Amplifier No.2 = 52.0 dB

Table 5-I lists the effect of two C/N values when being combined. For instance, our fiber transmitter C/N is 51 dB and the fiber cable C/N is 52.3 dB. The difference is 1.3 dB. The table shows the number 2.41 for 1.3 dB. 2.41 dB have to be subtracted from the lower value of 51 dB to arrive at the combined fiber transmitter plus fiber cable C/N. The result is 48.59 dB.

dB Difference	0	0.1	0.2	0.4	0.6	0.8
0	3.01	2.96	2.91	2.81	2.72	2.63
1	2.54	2.50	2.45	2.37	2.28	2.20
2	2.12	2.09	2.05	1.97	1.90	1.83
3	1.76	1.73	1.70	1.63	1.57	1.51
4	1.46	1.43	1.40	1.35	1.29	1.24
5	1.19	1.17	1.15	1.10	1.06	1.01
6	0.97	0.95	0.93	0.90	0.86	0.82
7	0.79	0.77	0.76	0.73	0.70	0.67
8	0.64	0.63	0.61	0.59	0.56	0.54
9	0.51	0.50	0.49	0.47	0.45	0.43
10	0.41	0.40	0.40	0.38	0.36	0.35

Table 5-I - Combining Two Carrier-to-Noise Ratios

By combining the two receiver C/N values, we obtain 51.0 -0.61 = 50.39 dB. Between the trunk station and one distribution amplifier, we obtain 50.24 dB. Combining the 48.59 dB of the transmitter and the fiber cable with the 50.39 dB for the two receiver values gives us the total C/N ratio of the fiber-optic segment. The result is 48.59 -2.2 = 46.39 dB.
Combining the coaxial broadband section, i.e. the 50.24 dB from above, with the second distribution amplifier gives us 50.24 -2.2 or 48.04 dB and the final calculation will combine the fiber link of 46.39 with the coaxial segment of 48.04 for a system C/N ratio of 44.11 dB.

Any C/N ratio exceeding 44 dB is considered quality video with no visual impairments. Please note that the HFC circuit in this calculation consists of a 26 km fiber cable connection. Your enterprise system will be substantially shorter and feature system carrier-to-noise ratios in excess of 45 dB.

To perform this calculation during the design phase of your HFC system, it is important that you obtain accurate information from the manufacturers of fiber-optic and coaxial equipment. Due to rapid innovations, the critical numbers that you need for your calculation are changing frequently.

The Composite Triple Beat

The composite triple beat or CTB is the second most important performance indicator. The higher the upper frequency of the HFC system and the higher the number of carriers are on the system, the higher the accumulation of third-order beats will be.

Fig. 5-3 shows the typical display of a TV channel at an RF frequency. Whether analog video or digital bit streams for voice, data or video, the 6 MHz regime remains the same.

A third-order intermodulation product contains the addition and subtractions of three carrier frequencies (F1 $\pm$F2 $\pm$F3). This third-order intermodulation product is often called triple beat.

In multichannel systems the limiting performance factor is usually the composite of all third-order beats. Experience shows that visual impairment of the picture quality becomes subjectively noticeable when CTB values are less than -48 dB below the carrier level. The total number of beats that pile up at a single frequency can be calculated. Manufacturer's specifications list CTB values for unmodulated carriers, which are at least 6 dB higher than a measured CTB level would be in the real world.

To minimize the CTB numbers, it has been the practice to harmonically generate all RF carriers. This is accomplished at the headend by locking all RF modulators to a comb or harmonic generator. This method will assure carriers to be evenly spaced, every 6 MHz, so that all beat products can only build up at the exact carrier frequencies. This keeps the beats out of the single-sideband area where the modulation resides.

Intermodulation products can impair the quality of a television picture ever so slightly. A large value of CTB products may appear as if the picture is a little out of focus or a little fuzzy and noisy. Managing CTB degradations in an HFC system is a high priority.

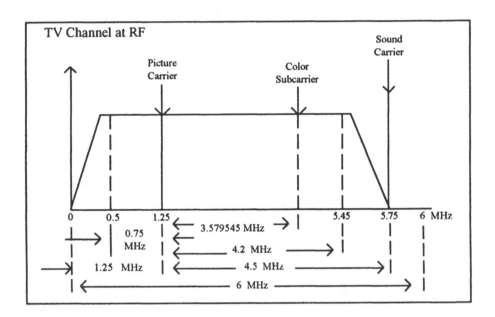

Channel Bandwidth at RF: 6 MHz
Visual Carrier Location: 1.25 MHz (± 1 KHz) above lower channel edge
Color Subcarrier Frequency: 3.579545 MHz (± 10 Hz) above visual carrier
Sound Carrier Center Frequency: 4.5 MHz (± 1kHz) above visual carrier
Scanning Lines: 525 lines per frame, interlaced 2-to-1
Scanning Sequence: Horizontally from left to right, vertically from top to bottom
Horizontal Scanning Frequency: 15,750 Hz (monochrome), 15,734.264 Hz (color)
Vertical Scanning Frequency: 80 Hz (monochrome), 59.94 Hz (color)
Aspect Ratio: 4 horizontal units, 3 vertical units

Fig. 5-3 A Television Channel at a RF frequency with a
channel bandwidth of 6 MHz

The Fiber-optic Segment

The fiber-optic segment can be designed for a fixed CTB value by adjusting the laser C/N ratio of the transmitter. For instance, the CTB value can be fixed at 65 dB and the transmitter C/N ratio is adjusted for channel loading. For a 60-channel loading, however, the C/N ratio cannot be increased over the 51 dB stated before. Nor can a CTB value be expected in excess of 62 dB.

The CTB performance of fiber-optic receivers is usually excellent because nonlinearities in the optical/electrical circuit can be kept to a minimum. Typical receiver CTB figures are in excess of lOO dB, even for a high number of channels.

The Coaxial Segment

This leaves us with the CTB performance of the broadband amplifiers. While trunk stations operating to 55O MHz feature CTB values of 7O to 9O dB, high-gain distribution amplifiers operating to 75O MHz with full channel loading have CTB values between 6O and 66 dB.

It is interesting to note that a high-split system operating between 222 and 750 MHz has about lO dB better CTB figures simply because all beat products below 22O MHz are not in the passband.

But for our calculation we will take the worst case of two distribution amplifiers operating in the sub-split configuration from 5O to 75O MHz. Again, we have to combine three different CTB values, the 62 dB for the fiber segment and two 66 dB numbers for the coaxial distribution amplifiers.

CTB is combined on a voltage basis, which at 6 dB per combining calculation has twice the effect of the C/N ratio.

Table 5-2 shows the values that need to be subtracted from the lower value in the combining process.

dB Difference	0	0.1	0.2	0.4	0.6	0.8
0	6.02	5.97	5.92	5.82	5.73	5.63
1	5.53	5.49	5.44	5.35	5.26	5.17
2	5.08	5.03	4.99	4.90	4.82	4.73
3	4.65	4.61	4.57	4.49	4.41	4.33
4	4.25	4.21	4.17	4.10	4.02	3.95
5	3.88	3.84	3.80	3.73	3.66	3.60
6	3.53	3.50	3.46	3.40	3.33	3.27
7	3.21	3.18	3.15	3.09	3.03	2.97
8	2.91	2.88	2.85	2.80	2.74	2.69
9	2.64	2.61	2.59	2.53	2.48	2.44
10	2.39	2.36	2.34	2.29	2.25	2.20

Table 5-2 - Voltage Addition Chart for combining two Composite Triple-Beat values

114

The two coaxial distribution amplifiers have a difference CTB of O. Therefore, we have to deduct 6.02 from 66 dB. The resulting 59.98 dB are subtracted from the 62 dB for the fiber segment, which gives us another 4.99 to deduct from 59.98 for a total system CTB of 55 dB. By subtracting another 6 dB from this number, to account for the modulation of all RF carriers, we have proven to ourselves that the composite triple beats of the system will be 49 dB below carrier levels.

Conclusions

By making a few essential calculations, you are now assured that the HFC system that you are planning for your enterprise will provide an excellent picture quality. The quality of performance of the system has been optimized and the availability of the system assures almost maintenance-free operation.

The calculations have proven that the fiber segment of the HFC system can be of considerable length to transport 60 plus video channels to a distribution area. The distribution area may consist of 2 to 3 amplifiers in cascade when planning to build a sub-split 50 to 750 MHz system. If your plans are leaning towards the high-split system concept, then 3 to 4 amplifiers in cascade can be used since channel-loading and triple-beat count is considerably less for the 222 to 750 MHz band.

In the return direction, i.e. for the sub-split band of 5 to 46 MHz or for the high-split return of 5 to 186 MHz no calculations are necessary. First, the channel-loading cannot exceed 30 channels. Second, the frequencies are lower and most triple-beat products fall out of band. Third, the C/N ratios and CTB numbers are inherently better at lower frequencies for both fiber and coaxial equipment.

Chapter 6

Planning the HFC Network

Inside-Plant Considerations

The Operations Center

The operations center is the heart of the HFC network.
The operations center consists of three major functions, i.e., communications management, circuit control/switching center and production.

The equipment complement depends upon the various functionalities that are initially selected as well as future growth requirements.

As the gateway to the outside world, the HFC operations center requires fairly easy access to outside communications facilities as well as physical co-location with the data network and its gateway requirements.

As the TV-channel selection and switching location, the operations center needs to be located close to satellite receiving antennas, uplink locations and off-air antenna installations.

As the netwok center, the operations center requires close proximity to the communications manager and to the administrative personnel of the enterprise.

Existing video production personnel will claim that proximity to existing studio facilities is absolutely essential to assure growth and to interface the production of video channels in an easy manner.

Finding the Location

It is recognized that the establishment of a location for the operations center is largely a political decision. Finding space for any new operation requires a careful search even in the smallest enterprise, school or hospital. The larger the enterprise, the bigger the campus and, therefore, the greater the involvement of company politics.

The ideal location of your future network operations center, of course, is

- next to the MDF for external communication facilities

- just below the roof that has already off-air antennas and satellite dishes

- next to the existing video production studio

- around the corner from the communications manager and the A/V facility

Obviously, there is no ideal location available that combines all of these desirable conditions. If space can be secured that meets the above ideal conditions, the installation will be most cost-effective.

That is not to say that video production facility could not be located in a different building. A separate interconnect cable and additional transmission equipment can solve the problem.

The same applies to off-air antennas and satellite up and downlink installations. These can be interconnected with the operations center over longer distances. However, separate interconnect cables and transmission equipment are required for distances in excess of 150 ft.

Locating the operations center must make sense from an operational and managerial point of view, but is largely depending upon the compromises that can be reached between interested parties.

It is good to remember, however, that the HFC operations center installation requires substantial entrance cabling to facilitate its functionality as the headend of the network. Small coaxial HFC networks only require a few large-size coaxial cables, but large HFC systems require multiple twelve-strand fiber-optic cables to each fiber-node and coaxial cables for local service.

This means that the HFC operations center should not be located in the top floor of a building unless a substantial riser stack is available to house the fiber-optic and coaxial cables emanating from it.

Equipment, Power and Space Considerations

The equipment complement can be installed in standard 19-inch equipment racks. The number of racks required is directly related to the operational functions:

- Off-air antenna and satellite receiving and processing equipment for 30 channels of TV can occupy 4 to 5 rack spaces

- Matrix switching equipment, VCRs, digital storage equipment and video retrieval or video-on-demand equipment can occupy another 5 to 6 rack spaces

- Studio equipment, whether linear or nonlinear, requires additional rack spaces and desktop terminal support

- The network management system requires space for a computer terminal and rack space for status monitoring equipment

- Depending on the size of the HFC network and the number of fiber-nodes, additional racks are required for fiber termination equipment

- Coaxial entrance equipment requires wall space to permit the installation of electronic and passive equipment on a wallboard

- In addition, there may be ATM switches and SONET multiplexers that may be required for campus interconnection or for common-carrier interfaces

Not just the initial space requirements need to be addressed, but any future expansion needs must be considered. As a result, the space assignment for the HFC operations center should be handled generously.

A minimum of 400 square feet is recommended. Additional space in the immediate vicinity is required for personnel, studio facilities, test equipment, bench-repair and network management, which may increase the space requirement to double or even triple that of the space requirement for the equipment.

The power requirement may suggest a small electrical room to house the various circuits. A 15 Amp. circuit is recommended for each equipment rack. Even though, video processing equipment uses less power and uses more efficient power supplies every year, the total power consumption of a video operations center equipment complement may approach 20 KW.

Fig. 6-I shows a typical layout of an operations center.

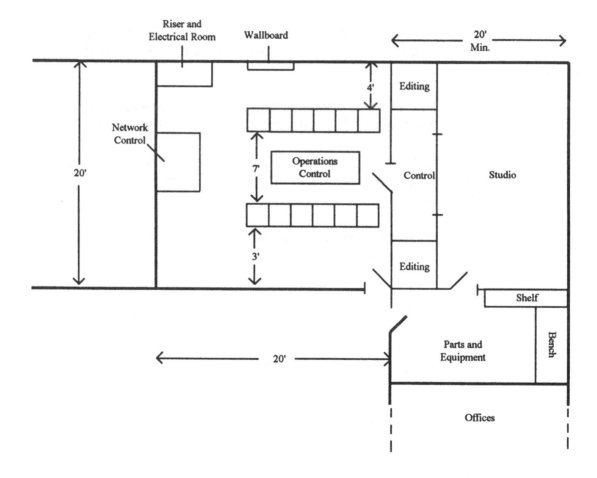

Fig. 6-1 Video Operations Center (typical)

Inside-Plant Data Collection

Whether your enterprise consists of one or dozens of buildings, the steps towards implementation of a broadband network are identical.

The starting point is to research the existence of CAD drawings of the campus geography with all buildings accurately dimensioned. The next step is to research the availability of floor plans of all buildings and to determine their accuracy and update status.

Accurate scale drawings of each floor of all buildings are one of the more important requirements in order to proceed with the planning of your HFC network. Floor plans marked not to scale are the next best thing. However, such buildings need to be surveyed and some measurements taken to establish a scale.

Take a look at the floor plans and see if they include a notation as to where the utility closets are located. If they are in the same position on all floors, you just found a riser. The floor plan might show the building-entrance location for the utilities.

- Are there room numbers?

- Do you maintain a room inventory system?

- Does your facility manager automatically enter new communication information?

- Are the telephone and data outlets that were installed last year included in the floor plans?

If none of this information is available, you have a difficult task ahead of you. If some of the information is missing, the planning process will take a little longer.

What is important in the planning phase is the accurate determination of cable routing, equipment placement and cable footage measurements. Without this data, the design of the HFC broadband network is not possible.

Building-Entry Locations

If your enterprise consists of a single building, the determination of a suitable building-entry location is of minor importance. Most likely, there is an electrical room or a telephone closet that contains cable terminations and a main distribution frame (MDF).

If you have your own private branch exchange (PBX), the equipment is probably located close to the MDF. The house wiring emanates from the MDF and goes to the various floors and to the outlets. Since telephone cabling has no distance limitation, older buildings may not feature risers and intermediate distribution frames (IDF) on every floor.

If your enterprise consists of a campus with multiple buildings, the location of the building-entry point is more important. The electrical room, depicted on your floor plan, may not be accurate and there may not be an existing riser.

A survey of each building is recommended to determine

- where a riser can be accommodated between the floors

and

- where the building entry location should be located to accommodate easy interface with both the prospective riser and the closest manhole

Small one or two-story buildings may not require any special considerations for the selection of a suitable building-entry location, especially when all service drops can be routed to the MDF building-entry location without exceeding the 150 ft. length limitation. Ideally, the MDF building-entry point is located in the center of the building to accommodate 150 ft. drops to either side on each floor.

Large horizontal buildings with only a few floors may require horizontal risers to accommodate the 150 ft. service drop limitation. In such a case, the building-entry location can be anywhere.

High-rise buildings usually feature riser rooms at each floor. The building-entry location logically coincides with the location of the riser stack.

MDF Mounting Considerations

Whether single-mode fiber-optic or coaxial cable comes into the MDF building entry room, there is a variety of equipment to be mounted:

- Fiber-optic receiver or coaxial distribution amplifier
- Splitters and test taps
- Multitaps for first floor service
- Power supply for fiber-optic node receiver
- Fiber-optic cable terminations
- Fiber-optic return transmitter
- Interconnect cabling
- Service drop terminations

This equipment can, of course, all be installed in a 19-inch standard equipment rack. However, since broadband coaxial components are designed for strand and pedestal mounting, an installation on a 3/4-inch plywood panel becomes easier, consumes less space and is more economical.

Fig. 6-2 shows a typical mounting arrangement on a 4x4 ft. wallboard.

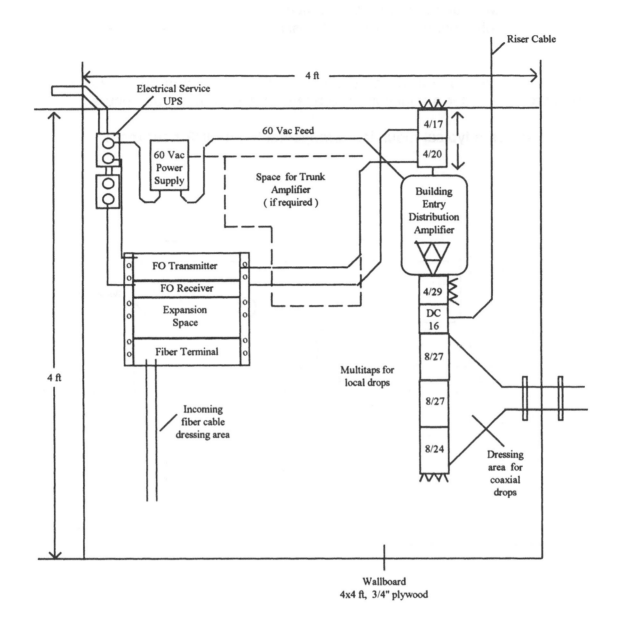

Fig. 6-2 MDF Typical Layout
Fiber Node Equipment and Building Entry Distribution Amplifier

IDF Mounting Considerations

The intermediate distribution frame (IDF) designation refers to the riser rooms on each floor of a building. As in the case of the MDF, a number of components have to be mounted at each floor level:

-       Splitters or directional couplers

- Distribution amplifiers - in case of a large building
- Multitaps to feed local service drops
- Interconnect cabling and possible horizontal riser cables

The mounting of these devices, again, is best accomplished on a plywood wallboard. Space for a minimum size board of 2 x2 ft. must be secured.

Fig. 6-3 shows a typical mounting arrangement in an IDF riser room.

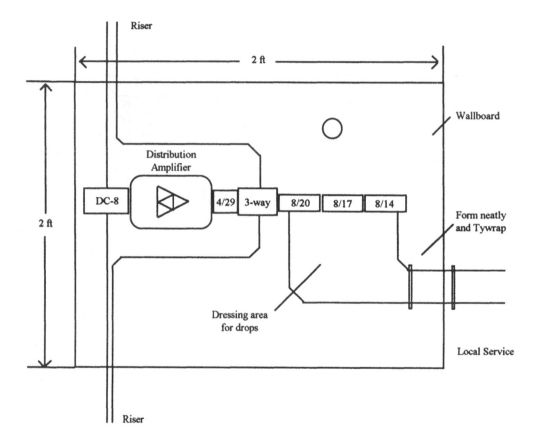

Fig. 6-3 IDF Typical Layout
Distribution Amplifier for service to upper and lower floors with 24 service drops

Locating the Broadband Outlets

The determination of the locations of all video outlets is the most important task of the planning process for a broadband video network.

The first activity in this process is a review of the floor plans and the preselection of rooms that require video or broadband services.

The next step is a physical survey of each preselected room to determine the desired outlet location(s). The following questions and answers may help to guide this planning process:

Q: *Should there be two outlets in a dormitory bedroom with two resident students?*

A: To install a splitter behind the faceplate is better than using an external splitter for the two TV sets. The number of maintenance calls will be reduced. The second outlet also may be used in the future for data or video transmission from a computer workstation.

Q: *Should the same outlet be used for data and voice?*

A: There is no reason for separating voice or data from the video. Modular single-gang faceplates are available for up to four different connectors. An RJ-II and an RJ-48 connector cable combined with two RJ-6 F-connectors on a single-gang faceplate. Dual-gang faceplates permit the installation of eight different connectors.

Q: *What is the recommended depth of the outlet box?*

A: A 2-inch depth is considered minimum to permit cable bending and proper terminating of the F-connectors. An additional miniature splitter can also be housed in a 2-inch deep housing. For additional data and voice terminations, a deeper housing is recommended.

Q: *Where does the outlet in a classroom go?*

A: If the TV set is ceiling or wall-mounted, the outlet is preferably installed at the level of the TV set.

Q: *Can I use the same outlet for a second TV set?*

A: It is expected that the second TV set is located across the room and that it requires its own faceplate and connector.

        In the case that the classroom will never be used for outgoing transmissions, a splitter could be located in the ceiling to serve both TV sets from the same service drop.

In the case of a transmit requirement, the first outlet would feature two F-connectors, one for transmit (T) and one for receive (R), with a splitter in the outlet housing. To split the signal again, for a second TV set, would reduce the minimum signal level below specification limits. It is, therefore, recommended to provide a second service drop for a second or third TV set.

Q: *Can a special level be provided for a large lecture room with six or eight TV sets?*

A: Yes, an 8-way splitter can be provided in the ceiling and connected to one service drop. However, this service drop would have to be fed from a special high-level multitap so that the outlet level for all outlets is maintained at least at +4.0 ±3.0 dBmV.

Q: *Where do I locate the outlet for a projection set?*

A: Every projection set has a control panel, which usually is wall-mounted. It is recommended to locate the outlet in the vicinity of the projection set's control panel.

Q: *What special wiring do I need for the retrieval of videos from a classroom?*

A: The classroom control unit is wired between the outlet and the TV set. There is a small fiber-optic wire that connects the classroom control unit with an IR transmitter that sits in front of the TV set's IR receiver window. If there is only one TV set, no additional wiring is required. If there are two or more TV sets, each must have such a fiber-optic wire to the IR window of the TV set. This means that IR wiring is required between the classroom control unit and each of the other TV sets. The wiring can be fed through a feed-through in the outlet faceplate over the ceiling and through another feed-through in the faceplate of the next TV set.

Q: *Do I need additional outlets to provide TV pictures to all computer terminals in a computer lab?*

A: To equip all computers with NTSC to VGA conversion, boards would be an expensive proposition. It is much more economical to connect the instructor's workstation with an NTSC to VGA conversion board and then use RGB wiring between computers and the instructor's workstation. Using a simple control panel, the instructor can now switch his own screen image (TV window, text or images) to all student workstation screens or to any screen that he may select. A video program, retrieved from the video operations center, can now appear on every screen in the computer lab. Also, any student, through the instructor's screen, can transmit his screen content converted to NTSC into the video network. No additional video outlets are required. However, daisy-chain cabling between all computer workstations and the instructor's console is necessary for the interconnection.

The 150 ft. Service Drop

Chapter 9 - The Network Design Process - deals with the calculation of attenuation at high and low frequencies. The reason for a length limitation of a service drop is directly related to the level difference between low and high frequencies and the higher attenuation of the small-size RG-6 service drop cable.

For instance, a 150 ft. drop using an RG-6 plenum-type cable has a difference of over 5 dBmV between 750 and 220 MHz. Between 750 and 50 MHz, the difference increases to over 7.5 dBmV. This means that the outlet level specification of +8.0 ±3.0 dBmV for 220 to 750 MHz service can barely be met with a 150 ft. drop length. The outlet level specification for 750 to 50 MHz must be relaxed to +10.0 ±4.0 dBmV as a minimum.

The 150 ft. service drop limitation and the recommendation to cut all drops to equal length is also directly related to the stringent level requirement for return transmissions. Varying drop lengths would cause the video level to arrive at the video operations center at very different levels and degrade the quality of the transmission. To control the level window of all return transmissions, the standard 150 ft. drop length is an absolutely necessary limitation.

To determine the riser interface location for the 150 ft. drop, it is important to measure the distance that the service drop can span. Assuming an 8 ft. vertical distance to the ceiling, the horizontal distance is 140ft. Physical measurements or scaling of the floor plan is required to confirm that the drop can be brought to the IDF or riser location on the same floor.

If this is not the case, measure and enter the distance between IDF location and the end of the 150 ft. drop on the floor plan and mark as horizontal riser

Fig. 6-4 shows a typical floor plan with 150 ft. drop sections and a horizontal riser.

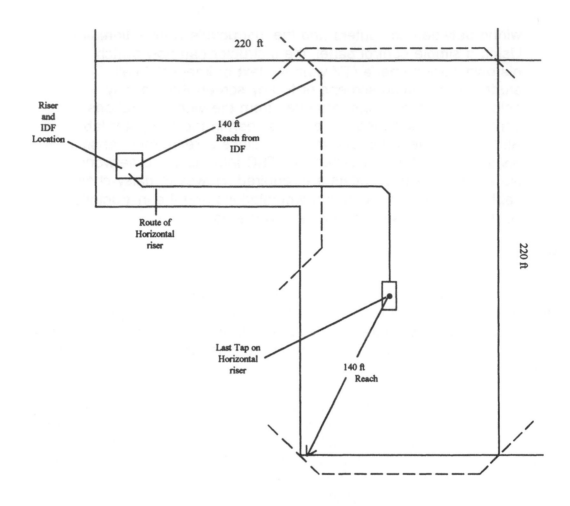

Fig. 6-4 Floor plan with Horizontal Riser to accommodate
drop lengths of 150 ft.

Riser Cable Considerations

The riser cable is the building distribution cable and extends from the MDF or building-entry location to all multitaps required to feed the service drops.

Riser cable is required to reach to all locations within I50 ft. from the selected outlets. When in a vertical orientation, the riser cable must be riser-rated and feature a flame-retardant polyethylene jacket. The riser rating must conform to National Electric Code (NEC), Article 82O. All cables used must be marked to identify the classification it is listed for. The reference marking for riser cable is 82O-5l (b).

Horizontal riser section can use riser-rated cables in nonplenum areas. When plenum conditions are encountered, plenum cable with a marking of 82O-5l (a) must be used.

Whether in vertical or horizontal position, the riser cable is the main distribution cable in a building.

Fig. 7-2 shows examples of riser networks. When amplification is required, it is important that secondary amplifiers are in a parallel configuration to optimize signal quality and minimize increases in noise caused by cascaded amplifiers.

The planning phase must collect all distances of the riser cable route. A physical survey and measurement of all distances from the MDF to the last multitap location is critical to assure the quality of the design process.

Outside-Plant Data Collection

Outside-plant data only needs to be collected for campuswide, privately owned enterprises.

It is hoped that facilities management has a CAD program and has been updating pole line and UG conduit networks to reflect current conditions. It is further desirable that the CAD drawings of the campus include manhole locations, conduit sizes and conduit occupancies.

If this information is not available at the outset of the planning phase for the video network, no harm is really done. The only problem is that more time will have to be spent to collect the information.

Even if the best campuswide CAD drawings exist and are updated to last week's status, there is information to be gathered relative to distances and available conduit space.

Aerial-Plant Data

Most pole lines in campus areas are in existence for many years and cluttered with multiples of electrical and communication wires.

A brief survey will show if another cable system can be accommodated. To facilitate the planning process, the distances between poles and the distances between the end poles and the adjacent manhole must be taken. A measuring wheel properly calibrated can be used to map the distances in an expedient manner.

Make-Ready Considerations

The distribution of the utility cables in the vertical plane is subject to joint-use pole make-ready regulations. The idea behind make-ready standards is to assure that communication facilities are never close enough to power lines to make contact.

Fig. 6-5 shows some of the more important clearance distances in a pole span. To confirm that a new cable system can be accommodated, it is recommended that vertical clearances be measured to assure minimum clearance requirements.

If the minimum clearances are not met, the pole line should be "made ready" before any additional cabling is placed.

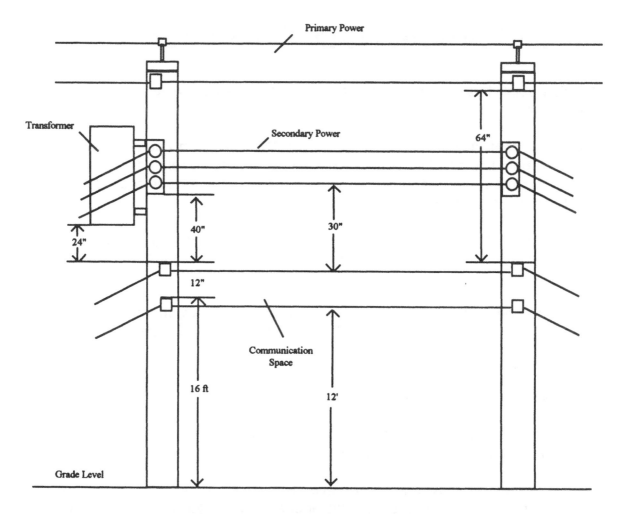

Fig. 6-5 Pole Line Data - Joint Use Agreement -
Separation of Power and Communications

UG Conduit Data Collection

Take a look at the campus CAD drawings. Are the building outlines correct? Are the buildings positioned correctly? Are the manholes located accurately? Is the drawing plotted to scale?

Distances between manholes are important for the planning process of the broadband HFC network. It may be much easier to send somebody out with a measuring wheel and enter these measurements on the drawing than to rely on the accuracy of the scale.

Manhole Locations and Sizes

The CAD drawing of your campus may show manholes and connecting lines between manholes. The CAD drawing may also identify electrical manholes and communication manholes. You are not interested in manholes used for power transmission. To combine telecommunications cabling with electrical cabling in one duct system is a dangerous procedure and defies electrical safety standards.

Whatever the review of existing CAD drawings may show, a survey of the existing conduit system is indicated. If the conduit plant has been installed during the last decade, there will be someone in your employment that knows exactly how the conduit was routed. In that case, measuring the distances between manholes, and between manholes and building entrance locations, will suffice. The measuring wheel has sufficient accuracy to compile all distances in an expedient manner.

If, on the other hand, the drawings are not to scale and the routing of the conduit is unknown, it may be more advantageous to conduct a more detailed survey. You need pullines for the new cable installation anyway. So the recommended procedure is to open the manholes, find a duct that is empty and insert a pulline with footage markers between manholes, to determine all distances accurately.

The survey also should evaluate the sizes of existing telecommunication manholes. The fiber and coaxial cables may have to be pulled through the manholes with vertical and horizontal displacements and angles, which requires room for handling of corner brackets and rollers.
Fig. 6-6 shows a typical conduit routing and manhole crossections.

Conduit Availability and Space Requirements

Opening the manholes and then comparing what you see with what the "updated" drawing says often leads to discouragement. You thought that there should be an unused 4-inch conduit. However, somebody pulled a 200-pair telephone cable through it.

> The diameter of a 12-strand armored single-mode fiber cable is about 0.35 inches

- The diameter of a 48-strand armored single-mode fiber cable is about 0.5 inches

- The diameter of a large coaxial trunk cable is about 0.87 inches

- The diameter of a typical coaxial distribution cable is about 0.5 inches

From this, we can see that even an empty 2-inch conduit can accommodate both fiber and coaxial cable types.

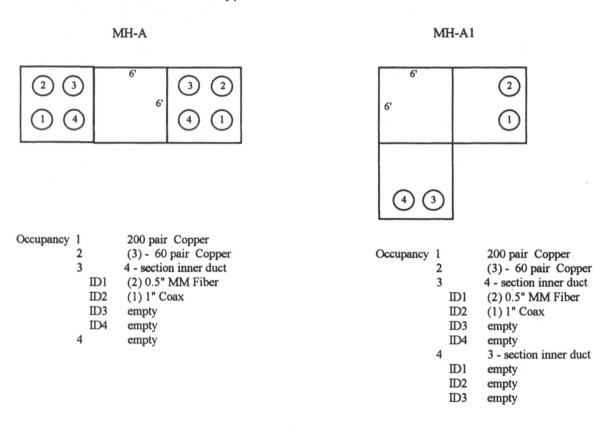

MH-A		
Occupancy 1		200 pair Copper
2		(3) - 60 pair Copper
3		4 - section inner duct
	ID1	(2) 0.5" MM Fiber
	ID2	(1) 1" Coax
	ID3	empty
	ID4	empty
4		empty

MH-A1		
Occupancy 1		200 pair Copper
2		(3) - 60 pair Copper
3		4 - section inner duct
	ID1	(2) 0.5" MM Fiber
	ID2	(1) 1" Coax
	ID3	empty
	ID4	empty
4		3 - section inner duct
	ID1	empty
	ID2	empty
	ID3	empty

Fig. 6-6 Typical Manhole Crossections and Occupancy

Under ideal circumstances, we are looking for an empty 4-inch conduit. Using a quad innerduct system, we would subdivide the 4-inch duct into four 2-inch innerduct segments, each equipped with pullines. Now, we have an infrastructure that can handle any required cable count that may be necessary in accordance with the completed design document.

Whatever the conditions of the existing duct plant may be, it is usually possible to pull an additional cable without too much of a problem. But, it is also recommended to prepare the conduit plant for the installation of an HFC network by

a)   rodding all desirable conduit segments
b)   installing a pulline with footage markers

In addition, a complete inventory of all telecommunication cable facilities may be taken. This is an opportune time to computerize existing plant records so that the new proposed HFC cabling information can be entered from the as-built records.

Cable Construction in Steam Tunnels

Many multi-building enterprises maintain a network of steam tunnels that are either vacated or have been elevated to public walkways that interconnect buildings.

A vacated steamduct can be used safely for HFC cable installation. The cable can be fastened to cable brackets without requiring the protection by a conduit.

Functioning steam tunnels require the careful planning of a conduit route. Engineering of a 4-inch conduit path is recommended. Care must be taken to bypass, cross or parallel existing conduits and steampipes. Detailed engineering documentation is required before commencing construction. Exact distances of any steam-tunnel installation must be collected to facilitate the network design process.

New Conduit Construction

The survey may reveal that connectivity for new cables cannot be provided in certain sections of the existing conduit network. Also, the selection of different building-entry locations may require new conduit paths.

In that particular case, it is recommended to proceed with the survey, engineering and implementation of the new conduit sections ahead of the HFC implementation so that this work does not affect the time schedule of the network implementation program.

Chapter 7

The Design Information Checklist

In-Building Information

Whether your enterprise consists of one building or groups of buildings, the development of a comprehensive planning document for the proposed broadband HFC video network is essential to prepare for the network design process.

What Information must be listed?

All information relative to the video network, i.e.

- floor plans indicating the physical location of all outlets

- riser-routing requirements

- measurement and reach of service drops

- room identifications with video usage information

- MDF and building-entry locations

What is the Information that should be listed in a service summary?

In order to be able to utilize the information in an effort to optimize the design in the most economical manner, it is important to summarize

- outlet totals per building and all buildings

- special horizontal riser requirements

- riser routing with measurements

Fig. 7-I shows a typical floor plan of floor #5 of the building No.I58 with all service drops located. Fig. 7-2 shows the vertical riser route from the MDF to the top floor with the horizontal riser extension on floor #5.

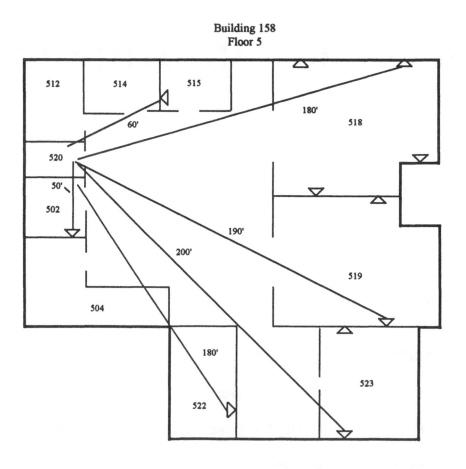

Fig. 7-1 Fifth Floor Floor plan - Building No. 158

Buildings, Floors, Rooms and Outlets

To collect all pertinent information of a particular building, a tabular listing, similar to that of Tables 7-I and 7-2 is recommended. Table 7-I simply summarizes the information that has been graphically shown in Fig. 7-I and 7-2.

Table 7-2 concerns itself with riser data and tabulates the graphic information provided in Fig. 7-2.

Building 158

Floor	Room No.	Usage	Video No. of Outlets	Data No. of Outlets	Service over 150 ft.	Status
1	101	Admin.	1	1	---	False Ceiling
1	104	"	1	1	---	"
1	105	"	1	1	---	"
1	108	"	1	1	---	"
1	115	Class	1	1	—	"
2	201	"	1	—	---	"
2	205	"	1	—	—	"
2	208	"	1	—	—	"
2	215	Lecture Hall	4	4	—	Molding
3	304	Class	2	2	—	False Ceiling
3	308	"	1	—	—	"
3	315	"	1	—	—	"
4	410	Admin	1	1	—	"
4	412	"	1	1	—	"
5	502	"	1	1	—	"
5	515	"	1	1	—	"
5	518	Lecture Hall	4	4	180	Plenum
5	519	"	2	2	190	Plenum
5	522	Computer Lab.	1	1		Plenum

5	523	"	2	2	200	Plenum
6	601	Admin	1	1	---	False Ceiling
6	605	"	1	1	---	"
Totals 6	Totals 22		Totals 31	Totals 26	Riser Floor 5	

Table 7-I - Listing of proposed video and data network outlets

This hypothetical example of building No. I58 provides information as to the total number of rooms requiring video service, multiple outlet requirements, room usage and the assessment of service drops in excess of I50 ft. In addition, the listed conditions provide information as to plenum cable requirements.

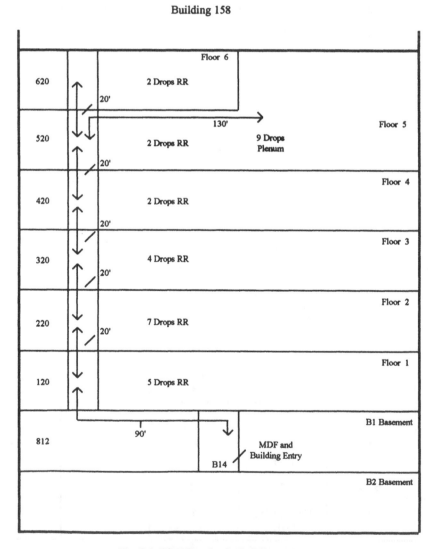

Fig. 7-2 Riser Routing in Building 158

Having recorded this information, we can now proceed to tabulate the riser route of Fig. 7-2. The important information relative to the riser are distances and routing details.

Building 158

Riser	Room	Footage to next Floor	Horizont Riser Extension	Wallboard {ft}	Riser Condition
MDF	B 14	90	---	4 by 4	Riser Rated
Floor 1	120	20	---	2 by 2	"
Floor 2	220	20	---	2 by 2	"
Floor 3	320	20	---	2 by 2	"
Floor 4	420	20	---	2 by 2	"
Floor 5	520	20	150	2 by 2	Plenum
Floor 6	620	---	---	2 by 2	Riser Rated
Totals		190 ft	150 ft Pl.	1 ea 4 by 4 6 ea 2 by 2	

Table 7-2 - Listing and Conditions of Riser Data

This is all the information on building No. l58 that we need in preparation for the design process. Although, we do not know yet what active electronics, passives and multitaps we need, we have the pertinent cable data such as

- footage of riser-rated riser cable

- footage of plenum-rated riser cable

- number of drops

- number of plenum drops

- footage of horizontal risers

- number of 2x2 ft. wallboards

- special wallboards, molding etc.

What you have compiled is the beginning of a comprehensive Bill-of-Material which, when supplemented with design data, gives you a very detailed picture of the in-building network.

Is this kind of Information required for every Building?

The answer is yes. Every building needs to be surveyed and the data compiled in a similar manner to Tables 7-I and 7-2. You may come across conditions that deserve attention and entry in your building record tables. There are a lot of opportunities for you to improve the layout of the tables. Your personal knowledge of the campus and your special desires can be incorporated best by yourself.

Upon collection of all individual building information, it is recommended to summarize all building information in a Summary of Buildings - Table 7-3. A hypothetical number of I5 buildings has been selected.

Building Summary

Item	Build No.	No. Floor	RR Riser	Hor Riser	Hor Plen. Riser	Video No of Rms.	Video Out lets	2by2 Wall Bds.	4by4 Wall Bds.	No of Drop	Plen. Drop
1	R100	16	350			192	192	16	1	192	
2	R110	12	280			192	192	12	1		192
3	120	3	80	240		12	18	2	1	18	
4	128	6	150			18	18	5	1	18	
5	132	8	180			24	30	7	1		30
6	140	3	80	180		12	18	2	1		18
7	148	3	80	240		18	18	2	1	18	
8	R150	8	180		600	180	180	7	1	180	
9	152	3	80			20	24	2	1	24	
10	158	6	190		150	22	29	5	1	26	3
11	R160	4	100			60	60	3	1	60	
12	R170	8	190			96	96	7	1		96
13	180	3	80	180		30	34	2	1	34	
14	182	3	80	180		30	34	2	1	34	
15	184	2		240		18	18	1	1	18	
Total			Total 2100	Total 1260	Total 750	Total 924	Total 961	Total 75	Total 15	Total 622	Total 339

Table 7-3 - The Building Summary Record

Supplementary Building Data

While surveying a building, it may be advantageous to record all information related to the installation of the video network. The data listed below is not pertinent to the network design process, but provides valuable information for RFP preparation and bidding effort.

Outlet Locations

-       Where are the exact outlet locations?
        Determine with horizontal and vertical measurements
        from corner and floor.

Faceplates

-       What kind of faceplate will be used?
        Is the outlet for video only or for type-5 data wiring as well?
        Determine your universal outlet arrangement that best fits
        the needs of the enterprise.

Wallboards

-       Where should the wallboards be mounted?
        Determine the exact locations of IDF and MDF wallboards
        by collecting horizontal and vertical measurements from
        convenient room references. Reserve the space using
        identification markers so that it is not blocked when you need it.

Horizontal Risers

-       Were should the horizontal riser be installed?
        Determine the route of the horizontal riser in the false ceiling
        area and determine minimum distances between fastening
        points. Describe the mounting method of the multitap(s) at
        the end of the horizontal riser.

Vertical Risers

-       Where should the vertical riser be installed?
        Determine the vertical routing from floor to floor and between
        the wallboards. Fastening of the cable at 6-ft. intervals is
        recommended. Check your fire regulations for fire stops
        between floors. Conduit raceways for vertical risers are not
        required for electrical reasons, but safety considerations and
        companywide universal wiring plans may dictate the ground rules.

Safety

- How do I secure MDF and IDF locations?
It is recommended that MDF building-entry closets and
telephone or IDF closets at floor levels are lockable.
A special universal lock and key should be used for all
telecommunication closets. The availability of a single
key reduces access problems during construction and
simplifies maintenance.

The 150-ft. Service Drop

- How do I install the 150-ft. service drop?
False-ceiling installation and fishing through the wall to
the outlet is recommended. Left-over cable shall be coiled
in the false ceiling above the vertical service section and
fastened using tie-wraps to an available structural member.
In the horizontal plane, service drops shall be bundled where
possible and connected to structural members using tie-wraps.

Installation in Molding

- How do I install service drops in hallways?
The use of suitable molding is the only way to protect the
cable and approach an aesthetic appearance. Select the
size of the molding for the largest crossection of cables
and carefully route and measure the proposed sections
and wall-penetration locations.

Outside-Plant Information

In an enterprise campus, the collection of pertinent outside-plant data is
essential to facilitate the network design process.

Fig. 7-3 shows a hypothetical campus layout, which consists of the 15 buildings
listed in the Building Summary of Table 7-3.

The campus layout of Fig. 7-3 is purely arbitrary but typical. The campus shows
a core group of buildings that mark the early or historical placement of
buildings. The earliest buildings are interconnected with an old pole line. When
buildings No. 128, 140 and 120 were added, they were connected by tunnels.
When building No. 158 was built, the first conduit section was constructed
consisting of MH-A and MH-A1. Then the residential buildings were built
together with building No. 184 and the UG conduit was extended to MH-E.

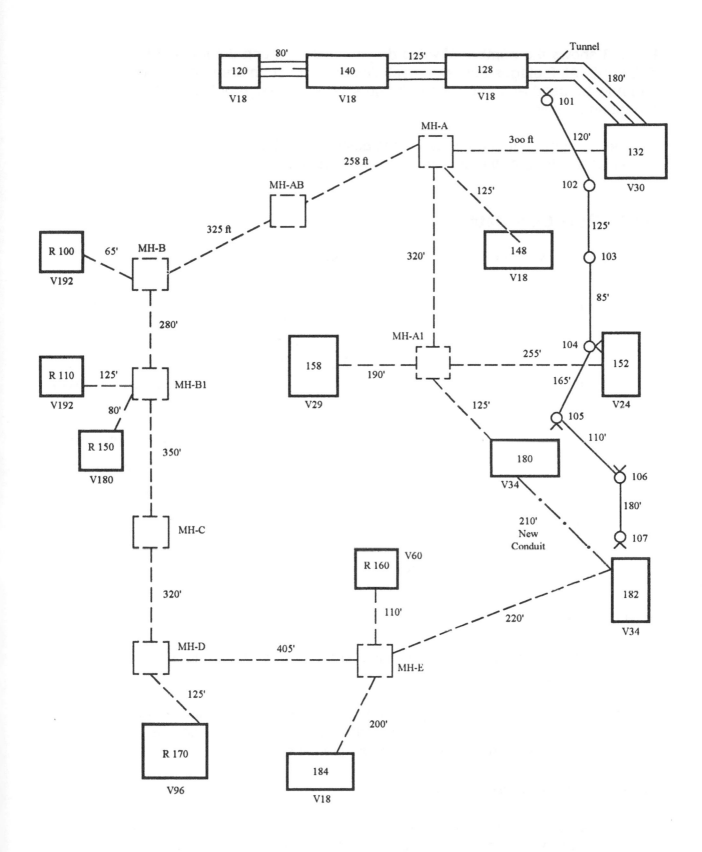

Fig. 7-3 The Hypothetical Campus Layout

Fig. 7-3 shows the number of video drops that have been determined from Table 7-3. All footages between manholes have been entered and the drawing can now be used as the basis for the outside-plant design of the HFC broadband network.

For the purpose of this application, it is assumed that the video operations center or headend of the system will be located in building No. l32.

Aerial Pole Line Data

The pole line data can be summarized in tabular form.

Pole No	Anchor	Distance from previous Pole	Comments
101	In line		clearence problem
102	---	120	"
103	---	125	21" to secondary
104	side	85	40" to primary
105	in line	165	12" midspan clearance
106	slack span	110	"
107	in line	180	"

Table 7-4 - Pole Line Data

A review of the conditions of the pole line shows that it should not be used for new cable installation.

There is, however, an interesting interconnect possibility. If we follow the conduit path to building No. l82, we find that there is no connection between building No. l82 and l8O except for the old pole line.

New Conduit Construction

Interconnection between building No. l82 and l8O is not important for video service or for the routing of HFC cables. However, when considering avoidance routing or SONET survivable ring architecture, a new conduit between building No. l8O and l82 would be recommended.

144

Existing Conduit Data

Fig. 7-3 shows the various conduit sections and the distances between manholes and buildings. In order to prepare for the HFC network design process, all conduit data must be tabulated. Using building No. I32, the location of the video operations center as the starting point, distances are accumulated and compiled for each building location.

Table 7-5 shows the existing conduit data together with occupancy and space problems.

From Building	To Building	Distances {ft}	Cum Distances {ft}	Conduit Availability
132	128	180	180	Tunnel
128	128	250		Basement
128	140	125	555	Tunnel
140	140	250		Basement
140	120	80	885	Tunnel
132	MH-A	300		1-4"OD
MH-A	148	125	425	1-2"OD
MH-A	MH-A1	320		1-4"OD
MH-A1	152	255	875	1-2"OD
MH-A1	180	125	745	1-2"OD
MH-A1	158	190	810	1-4"OD
MH-A	MH-AB	258		2-4"OD
MH-AB	MH-B	325		2-4"OD
MH-B	R100	65	948	1-2"OD
MH-B	MH-B1	280		2-4"OD
MH-B1	R110	125	1288	1-2"OD
MH-B1	R150	80	1243	1-2"OD
MH-B1	MH-C	350		1-4"OD
MH-C	MH-D	320		1-4"OD
MH-D	R170	125	1958	1-4"OD
MH-D	MH-E	405		1-4"OD

MH-E	184	200	2438	1-2"OD
MH-E	R160	110	2348	1-2"OD
MH-E	182	220	2458	1-2"OD
182	180	210		New 2-4"OD
Total		5273		

Table 7-5 - Conduit Data

The table shows a total conduit footage of 5,273 ft. The portion of the conduit that is required in tunnels and in buildings is 885 ft. The total footage of the UG conduit is 4,178 ft. and new conduit is required for 210 ft.

The column entitled "Cumulative Distances" indicates the footages between building-entry locations and building No. 132, the headend location.

The HFC Criteria

At this point in time, we have compiled a complete record of all information pertinent to the design of the video network. We know footages to individual buildings. We know the number of video service locations in each building.

What are the conditions that affect the decision to use fiber-optic cable vs. coaxial cable?

There are a number of considerations that can help to determine the fiber/coaxial interface locations:

- a fiber-node should be located strategically to serve not more than 500 outlets and not less than 150 outlets

- the coaxial plant should not utilize more than 4 amplifiers in cascade when connected to a fiber-node

When we analyze the hypothetical campus layout of Fig. 7-3, we find that buildings R100, R110 and R150 together serve 564 outlets and each building meets the lower limit of 150 outlets.

Based on the footage count of Table 7-5, the longest distance between the headend in building No. 132 and the last building No. 182 is 2,458 ft. Using heavy 860 coaxial cable, the distance between amplifiers at 750 MHz is limited to 1,600 ft. Since two amplifiers would be required in the outside-plant portion and two amplifiers in the in-building portion, the limits for a solely coaxial system are exceeded or only marginally met.

However, if we change the routing and serve R170 through the new conduit between buildings No. 182 and 180, the longest distance has been reduced to 1,705 ft., which may reduce the amplifier cascade to the marginal number of 4.

HFC Alternatives

Now, that we have determined that a purely coaxial system would be marginal - what alternatives are left for consideration?

Fiber to all Buildings

The decision to use fiber in the outside plant and coaxial cable for in-building distribution is an easy one to make for many reasons:

- it is politically correct

- it is a nontechnical decision

- it is fashionable

- fiber is the future

But the negative factors outweigh the easy decision-making:

- many fiber cables are required

- more duct space is required

- it is less economical

- higher expenses for return-transmission

- more power-supply locations

To illustrate just the point about the duct space. Let us assume that we want to use 12-strand fiber cables for each building; in the crossection between building No. 132 and MH-A we would have to install eleven cables with 12 strands each. The conduit space required for this massive cable installation would be in excess of the one 4-inch OD duct that is available and new duct construction becomes necessary.

Fiber-optic cables to all building-entry locations is expensive, wasteful and technically unjustified.

The Optimized HFC System

The optimization process of an HFC transmission system is primarily a function of the number of users and the distance experienced in a group of buildings. Using common sense as the main criteria, the optimized HFC system for our hypothetical campus of Fig. 7-3 may look as follows:

a) Coaxial service for building No. 132 and buildings No. 128, 140, 120 and 148. These buildings are adjacent to the video operations center and have only a total of 102 outlets. No more than one amplifier is used per building. All amplifiers can be in parallel

b) A fiber-node for building R100 with 192 outlets, requiring 2 amplifiers in cascade within the building

c) A fiber-node for building R110 with 192 outlets, requiring 2 amplifiers in cascade within the building

d) A fiber-node for building R150 with 180 outlets, requiring 2 amplifiers in cascade within the building

e} A fiber-node for building No. 158 with coaxial distribution for buildings No. 152 and 180. There will be only 1 amplifier required per building and all amplifiers can be in parallel. There are only 87 outlets. If the new conduit is constructed, building No. 182 should be included in this node

f) A fiber-node for building R170 with coaxial distribution for buildings R160, R170, 184 and possibly 182. This node would consist of a maximum of 208 outlets. The coaxial plant in R170 would consist of 2 amplifiers in cascade. No more than 2 amplifiers in cascade would be required for buildings R160, 184 and 182

This optimized HFC system would use 5 fiber-nodes. The crossection of the cable requirement between building No. 132 and MH-A has been reduced to five 12-strand fiber-optic cables and one coaxial cable. The longest fiber-cable distance is to building R170 for 1,958 ft.

Fiber-optic cables are used between

- Building No. 132 and R100
- Building No. 132 and R110

- Building No. 132 and R150
- Building No. 132 and R170
- Building No. 132 and 158

Coaxial cables are used between

- Building No. 132 to 128, to 140, to 120
- Building No. 132 and 148
- Building No. 158 to MH-A, to 152 and 180
- Building R170 to MH-E and to R160, 182 and 184

The architecture of the optimized HFC system is considered preliminary and subject to design confirmation. Chapters 9 and 10 entitled "The Video Network Design Process" deals with detailed calculations to confirm or change this selection of fiber-nodes.

It is important to note that the review of outlets, in-building and outside-plant distances has given the designer the ability to rationalize, to optimize system performance, to determine the most economical solutions and to determine a fiber and coaxial routing plan that fits the geography of our hypothetical campus quite well.

Chapter 8

The HFC Network
Components

The components of the broadband networks are listed in the following with their respective electrical and mechanical specifications. The listing only includes infrastructure equipment required for the transmission of broadband signals inclusive of RF modulators and demodulators.

To utilize the capacity of the HFC broadband transmission systems, a number of companies are developing RF modems for telephony, ethernet, T-I as well as for MPEG -I.5, MPEG-2 and HDTV formats. Until these developments are completed and fully tested in the HFC environment, a treatment of electrical specifications is considered premature.

Fiber-Optic Cable and Equipment

Single-Mode Fiber-Optic Cables

Common Specifications

There are three types of fiber-optic cables - tight buffer, breakout and loose buffer.

The electrical specifications are identical for each of the above types:

Wavelength	1310nm/1550nm
Attenuation (dB/Km)	0.35/0.25
Core Diameter (microns)	8.3 nominal
Cladding Diameter (microns)	125 $\pm$1.0
Coating Diameter (microns)	245 $\pm$10.0
Cut-off Wavelength (nm)	1250 $\pm$70
Zero Dispersion Wavelength (nm)	1310 $\pm$10

The differences are predominant in the mechanical makeup of the cables.

The optical fiber is a very small waveguide. In an environment free from stress or external forces, this waveguide will transmit the light launched into it with minimal loss or attenuation.

To isolate the fiber from these external forces, two first-level protections of fiber have been developed: loose buffer and tight buffer.

Once a tight or loose buffer construction is selected, the system designer has made some decisions regarding the tradeoffs between microbending loss and flexibility in obtaining his optical operation goals.

For installation of cable, mechanical properties such as tensile strength, impact resistance, flexing and bending are important. Environmental requirements concern the resistance to moisture, chemicals and other types of atmospheric or in-ground conditions.

Normal cable loads sustained during installation may ultimately place the fiber in a state of tensile stress.

The level of stress may cause microbending losses, which result in an attenuation increase and possible fatigue effects.

To transfer these stress loads in short-term installation and long-term application, various internal strength members are added to the optical cable structure.

Such strength members provide the tensile load properties similar to electronic cables and keep the fibers free from stress by minimizing elongation and contraction. In some cases, they also act as temperature stabilization elements.

Optical fiber stretches very little before breaking, so the strength members must have low elongation at the expected tensile loads.

The Loose-Tube Cable

In the loose buffer construction, the fiber is contained in a plastic tube that has an inner diameter considerably larger than the fiber itself. The interior of the plastic tube is usually filled with a gel material.

The loose tube isolates the fiber from the exterior mechanical forces acting on a cable. For multifiber cables, a number of these tubes, each containing single or multiple fibers, are combined with strength members to keep the fibers free of stress and to minimize elongation and contraction.

By varying the amount of fiber inside the tube during the cabling process, the degree of shrinkage due to temperature variation can be controlled and, therefore, the degree of attenuation over a temperature range is minimized.

Loose-tube cable usually features 6 or 12 fiber strands in a bundle within a buffer tube.

The loose-tube cable construction utilizes

- a filling compound within the buffer tube
- a flooding compound between the buffer tubes
- blank fillers to substitute for not required buffer tubes
- an aramid yarn layer to increase the tensile strength
- an armor (if required)
- a riser-rated jacket

The makeup of loose-tube cable comes for various fiber count numbers. Since we are not building the information superhighway, only two values of fiber counts are shown:

Fiber Count	2-36	37-22
Number of Buffer Tubes	6	6
Outer Diameter	.56 in.	.56 in.
Weight (lbs./1000 ft.)	150	150
Max. Tensile Load during installation	600 lbs.	600 lbs.

during operation	200 lbs.	200 lbs.
Min. Bending Radius		
under load	II in.	II in.

All loose tube cables come in lengths up to 35,000 ft. on reel sizes from 42-inch diameter for 5,000 ft. to 78-inch diameter for 35,000 ft. The weight of the cable is approximately 150 lbs. per 1,000 ft. Armored cable double the crush resistance from 500 to 1,000 lbs.

Each construction has inherent advantages. The loose buffer tube offers lower cable attenuation from microbending in any given fiber, plus a high level of isolation from external forces. Under continuous mechanical stress, the loose tube permits more stable transmission characteristics.

The Tight Buffer Cable

The tight buffer uses a direct extrusion of plastic over the basic fiber coating.

Tight buffer constructions are able to withstand much greater crush and impact forces without fiber breakage.

The tight buffer design, however, results in lower isolation for the fiber from the stresses of temperature variations. While relatively more flexible than loose buffer, if the tight buffer is deployed with sharp bends or twists, optical losses are likely to exceed nominal specifications due to microbending.

The makeup of the tight buffer cable, again, features 6 or 12 fiber strands in a bundle:

- binder threads around each bundle
- a filling compound between bundles
- a large buffer tube combining all bundles
- a moisture barrier and an outer jacket
  with a dielectric strength member

The armored version features an armor inside the outer jacket and may contain a steel strength member within the outer jacket.

Because of the makeup, the cable counts are somewhat different:

Fiber Count	2-48	49-96
Number of Buffer Tubes	I	I
Outer Diameter	.49 in.	.57 in.
Weight (lbs./1000 ft.)	115	152
Max. Tensile Load		
during installation	600 lbs.	600 lbs.
during operation	200 lbs.	200 lbs.

Min. Bending Radius		
under load	lO in.	ll.5 in.

The crush resistance for the armored cable, again, is l,OOO lbs. and twice that of the nonarmored variety.

The Breakout Cable

A refined form of tight buffer construction is breakout cable. In breakout cable, a tightly buffered fiber is surrounded by aramid yarn and a jacket, typically PVC. These single-fiber subunit elements are then covered by a common sheath to form the breakout cable. This "cable within a cable" offers the advantage of direct, simplified connector attachment and installation.

Breakout cables are light-weight and more flexible. The fiber strands are not bundled but reside in the center of a buffer tube. The fiber strands are kept in the center of the buffer tube with a filling compound.

These cables are available of smaller fiber counts:

Fiber Count	2-l2
Number of Buffer Tubes	l
Outer Diameter	.35 inc.
Weight (lbs./lOOO ft.)	57
Max. Tensile Load	
during installation	4OO lbs.
during operation	9O lbs.
Min. Bending Radius	
under load	7.O in.

This cable is usually also available in an armored version, which, again, doubles the crush resistance from 5OO to l,OOO lbs.

Fiber-Optic Termination Equipment

Distribution and Storage Panels

There are many suppliers of rack-mountable fiber termination and storage equipment. New versions come on the market almost every day.

It is important that fiber terminations and connector panels incorporate the following parameters:

- rack- and wall-mountability
- minimum rack-space requirement

- grounding provisions
- strain relief
- storage for excess fiber
- convenience for splicing of trunk cable to pigtails
- protective storage for all splices
- independent splice drawers each with one or two splice trays
- removable front panel and fold-down polycarbonate front door for easy viewing of patch-cord connections
- complete with single-mode SC or ST adapters and receptacles

Single-mode Patch Cords

Single-mode patch cords require ultra PC polish to keep return loss as high as possible and the insertion loss as low as attainable.

Patch cords are usually available in lengths of 3 or 6 meters. The connector style shall be ultra PC SC or ultra PC ST.

The return loss of the patch cord must be greater than 50 dB and should be coded by the manufacturer to track the process history of the cord.

Dual-cable patch cords may offer an added installation ease. The insertion loss of the assembly shall not exceed 0.2 dB.

Single-mode Fiber Connectors and Accessories

Connectors shall be ST or SC with bayonet or push-pull housing. The insertion loss shall be 0.2 dB or better.

The back reflection shall be better than -35 dB (0.1 to 0.01%) as obtained with PC polish. Super PC polish is recommended.

The use of duplex coupling receptacles or duplex clips for combining two single connectors is recommended.

The light crimp ST connector offers an epoxyless termination to be completed in less than 2 minutes. It consists of a two-step crimping action to secure the fiber, cleave and 30-second polish.

Bulkhead adapters, if required, can be used but shall be ST or SC with zirconia sleeve. When using bulkhead adapters in fiber wall-mount boxes, use plastic angled retainers that can be individually removed for easy connector cleaning.

ST build-out receptacles are used in conjunction with ST build-out attenuators to reduce optical power in an optical network. The ST build-out receptacle can be mounted in the ST angled retainer.

ST build-out attenuators are used in conjunction with ST build-out receptacles to reduce optical power in an optical network. ST build-out attenuators snap into the build-out receptacle and are designed to operate at either 1310 nm or 1550 nm levels.

Build-out attenuators are available in 5 dB steps from 0 to 20 dB.

Fiber-Optic Transmission Equipment

Single-Channel Transceivers

Single-channel fiber-optic transceivers are baseband units good for only one channel of full-motion video and audio. While they are not used in multichannel HFC transmission systems, they may be an inexpensive solution for a single-channel return signal from an isolated low-density building.

The specifications feature:

Video Input/Output	1 Vp-p
Return Loss	30 dB
Bandwidth	20 Hz to 10 MHz
Signal-to-Noise Ratio	62 dB
Transmitter Power	1000 µW
Receiver Sensitivity	50 µW

The optical budget is usually around 10 dB, which permits distances up to 20 km. Rack-mountable shelf designs are available for central locations for baseband connection to matrix routing switches. Wall-mountable transceivers are available for easy installation at remote locations. To translate the baseband signals to an RF channel assignment, a standard modulator can be used. A demodulator is required to take the signal from the RF frequency assignment to baseband for transmission on the fiber.

Multichannel Transmitters

Designed for operation on multichannel HFC systems, there are a wide variety of low-power and high-power fiber-optic transmitters on the market, while transmitters at central locations are usually rack-mountable together with receivers and RF amplifiers. The horizontal layout in a 1 RU profile is also common. The transmission equipment for remote locations can have various types of enclosures. Examples are a single RU profile, rack-mountable, upright units or outdoor cast aluminum housings for combining the return transmitter with the forward receiver and RF amplifier.

While outbound fiber-optic transmitters are required to transmit a band as wide as possible (50 to 550 and 50 to 750 MHz), the inbound or return transmitters are usually low-powered and designed for 5 to 200 MHz transmission.

The typical specifications for forward transmitters are provided below:

	Medium Power	High Power	Low Power	Low Power
Frequency Response (MHz)	50-750	50-750	5-550	5-550
Wavelength (nm)	1310	1310	1310	1310
Transmit Power (mW)	9	12	.25	.3
(dBµW)	39	42.5	22.5	24.5
Optical Budget	11	12.5	9.5	11.5
Impedance (ohm)	75	75	75	75
RF Input Level (dBmV)	32	32	32	32
Flatness (dB) 50-750 MHz	±1.5	±1.5	±1.0	±1.0
C/N Ratio (dB)	51	52	50	51
Composite Second Order Beat (dB)	-60	-60	-60	-60
Composite Triple Beat (dB)	-65	-65	-62	-62.5
Return Loss (dB)	16	16	16	16
Supply Voltage (Vac)	110	110	110	110
Current (Amps)	0.5	0.8	0.2	0.25

In comparison, the specifications for return transmitters are usually based on a 5 to 200 MHz return-band. This 30-channel loading requires less stringent specifications and can be provided by less costly Fabry Perot laser equipment.

	Fabry Perot	Fabry Perot	Low Power	Medium Power
Frequency Range	5-186	5-200	5-200	5-200
Channel Loading	8	24	32	32
Transmit Power (dBµW/ch)	23	23	22	35.5
Optical Budget	10.0	10.0	9.0	11.0
RF Input Level (dBmV)	32	32	32	32
Impedance (ohm)	75	75	75	75
Flatness (dB) 5-200 MHz	±1.0	±1.0	±1.0	±1.0
Composite Triple Beat (dB)	-62.5	-62.5	-62	-65
C/N Ratio (dB)	50	50	51	51
Return Loss (dB)	16	16	16	16

Even if you are planning for a sub-split system with a 5 to 46 MHz return, the above fiber-optic return transmitters have to be used. There is no 5 to 46 MHz transmission equipment. The HFC philosophy is to convert a number of 5 to 46 MHz return ranges into a 5 to 200 return-band for the fiber-optic return segment.

Multichannel Receiver

There is a wide variety of fiber-optic receivers on the market that are all based on the principle of operation of the photodiode. The optical signal is received, translated to RF bandwidth and amplified to the required output level.

Receivers for headend and central location installation can be either obtained in vertical chassis packages or self-contained I RU rack-mountable units. At the remote location, the receiver can either be mounted above the fiber terminal on a rack-mountable wallbracket or may be in an outdoor amplifier housing suitable for mounting on a wallboard.

Receivers are usually designed for bandwidth so that they can be used in both forward and return directions.

The typical specifications are outlined below:

Frequency Range (MHz)	5-200	5-550	5-750
Channel Capacity	24-32	32-80	60-80
RF Output Level (dBmV)	+13.0	+13.0 or 33.0	+33.0
Optical Return Loss (dB)	45	40	40
Flatness (dB)	$\pm$0.5	$\pm$0.5	$\pm$0.5
C/N Ratio (dB)	53.5	51.5	51.0
Composite Triple Beat(dB)	122.0(24)	108.0(60)	80.0(80)

Fault Alarm and Telemetry

When purchasing fiber-optic transmitters and receivers, it is important to consider the ability of the units to report transmission failures.

Fault alarm and telemetry reporting of the operational integrity of the units will greatly assist in the management of the network.

The vital signs of a transmitter include
- loss of power
- loss of signal
- the laser temperature/Bias

The vital signs of a receiver include
- loss of power
- loss of signal
- low received optical power

The telemetry section of the unit must be wired in a manner that will combine the fault alarm conditions of muiltiple transmitter and receiver units.

Coaxial Cable and Equipment

Coaxial Cables

The important properties of HFC coaxial cables are indicated in this section. Each cable type is shown with its inherent mechanical composition and attenuation data.

Outside-Plant Cables

A wide range of outside cables exist. The main differences are the cable sizes. Larger diameters have lower attenuation. The type of dielectric used also changes the attenuation. The more air in the cellular structure of the dielectric, the lower the attenuation.

The following specifications cover outside-plant cables using an expanded polyethylene dielectric and a medium density polyethylene jacket.

Standard Dielectric Cables

Physical Demensions	500 Series inches	625 Series inches	750 Series inches	875 Series inches
Nominal Center Conductor Diameter	0.109	0.137	0.167	0.194
Nominal Diameter Over Dielectric	0.450	0.563	0.678	0.797
Nominal Diameter Over Outer Conductor	0.500	0.625	0.750	0.875
Nominal Outer Conductor Thickness	0.025	0.031	0.036	0.039
Jacket Versions				
Nominal Diameter Over Jacket	0.560	0.685	0.820	0.945
Nominal Jacket Wall Thickness	0.030	0.030	0.030	0.030
Underground (JCASS) Version				
Nominal Diameter Over Jacket	0.570	0.695	0.830	0.955
Armored Versions				
Nominal Diameter Over Corrugated Armor	0.635	0.755	0.920	1.017
Nominal Sheild Thickness	0.008	0.008	0.008	0.008
Nominal Diameter Over Outer Jacket	0.715	0.835	1.000	1.097
Nominal Thickness of Outer Jacket	0.040	0.040	0.040	0.040

The differences in the mechanical performance of these cables are as follows:

	500 Series	625 Series	750 Series	875 Series
Mechanical Characteristics				
Minimum Bending Radius:				
(No Jacket)	6.5 in.	7.5 in.	9.0 in.	10.0 in.
(Jacketed)	6.0 in.	7.0 in.	8.0 in.	9.0 in.
(Armored)	8.5 in.	9.5 in.	10.5 in.	11.5 in.
Maximum Pulling Tension:	300 lbs	475 lbs	675 lbs	875 lbs
Maximum D.C. Resistance @ 68° f. (20° C.)				
Copper Clad				
(Inner Conductor) / 1000 ft.	1.35 ohms	0.84 ohms	0.57 ohms	0.42ohms
(Outer Conductor) / 1000 ft.	0.37 ohms	0.23 ohms	0.19 ohms	0.55 ohms
(Loop) / 1000 ft.	1.72 ohms	1.07 ohms	0.76 ohms	0.55 ohms

The attenuation over frequency of these cables is compared in the following:

Frequency (MHz)	500 Series Maximum dB/100'	625 Series Maximum dB/100'	750 Series Maximum dB/100'	875 Series Maximum dB/100'
5	0.16	0.13	0.11	0.09
30	0.40	0.32	0.26	0.23
45	0.49	0.41	0.33	0.28
50	0.52	0.42	0.35	0.30
55 (Ch. 2)	0.54	0.46	0.37	0.33
83 (Ch. 6)	0.66	0.57	0.46	0.41
108	0.75	0.63	0.52	0.45
150	0.90	0.77	0.62	0.55
181	1.00	0.85	0.68	0.60
193	1.03	0.88	0.71	0.62
211 (Ch. 13)	1.09	0.92	0.74	0.66
220	1.11	0.94	0.76	0.67
250	1.20	1.00	0.81	0.72
270	1.24	1.02	0.84	0.73
300	1.31	1.04	0.89	0.78
325	1.37	1.13	0.93	0.81
350	1.43	1.18	0.97	0.84
375	1.47	1.22	1.01	0.88
400	1.53	1.27	1.05	0.91
425	1.57	1.32	1.08	0.95
450	1.63	1.35	1.12	0.97
500	1.73	1.43	1.18	1.03
550	1.82	1.50	1.24	1.08
600	1.91	1.58	1.31	1.14
750	2.16	1.78	1.48	1.29
865	2.34	1.93	1.61	1.41
1000	2.52	2.07	1.74	1.53

As can be seen, while the differences are small at low frequencies, the difference between the 500 and the 875 series cable at 750 MHz is almost 1dB per 100 ft at 1000 MHz.

Special Dielectric Cables

Special dielectrics with larger but uniform air pockets decreases the attenuation. As examples, 540 and 860 series cables are compared:

Physical Demensions Component	540 Series Inches	860 Series Inches
Nominal Center Conductor Diameter	0.124	0.203
Nominal Diameter Over Dielectric	0.513	0.828
Nominal Diameter Over Outer Conductor	0.540	0.860
Nominal Outer Conductor Thickness	0.0135	0.016
Jacket Versions		
Nominal Diameter Over Jacket	0.610	0.960
Nominal Jacket Wall Thickness	0.035	0.050
Armored Versions		
Nominal Diameter Over Corrugated Armor	0.685	1.030
Nominal Sheild Thickness	0.008	0.008
Nominal Diameter Over Outer Jacket	0.765	1.110
Nominal Thickness of Outer Jacket	0.040	0.040
Messenger Version		
Diameter of Steel Messenger	0.109	0.188

The mechanical differences and DC resistances are compared below:

Mechanical Characteristics	540 Series	860 Series
Minimum Bending Radius:		
(Jacketed)	4.0 in.	7.0 in.
(Armored)	6.5 in.	9.5 in.
Maximum Pulling Tension	220 lbs	450 lbs
Minimum Breaking Strength of Messenger	1,800 lbs	3,900 lbs
Maximum D.C. Resistance @ 68° f. (20° C.)		
(Inner Conductor) / 1000 ft	1.02 ohms	0.406 ohms
(Outer Conductor) / 1000 ft	0.59 ohms	0.318 ohms
(Loop) / 1000 ft	1.61 ohms	0.724 ohms

A comparison of attenuation over frequency shows that the 860 series cable has about 0.25 dB lower attenuation than the larger 875 series cable at 750 MHz.

Attenuation [@ 68° F. (20° C.)] Frequency (MHz)	540 Series Maximum dB/100'	860 Series Maximum dB/100'
5	0.14	0.09
30	0.34	0.23
45	0.41	0.29
50	0.44	0.30
55 (Ch .2)	0.47	0.32
83 (Ch. 6)	0.58	0.40
108	0.66	0.45
150	0.79	0.54
181	0.88	0.59
193	0.90	0.60
211 (Ch. 13)	0.95	0.64
220	0.98	0.65
250	1.03	0.70
270	1.07	0.72
300	1.13	0.76
325	1.18	0.80
350	1.23	0.83
375	1.27	0.86
400	1.32	0.88
425	1.37	0.92
450	1.40	0.95
500	1.49	1.00
550	1.56	1.06
600	1.64	1.10
750	1.85	1.24
865	2.00	1.33
1000	2.17	1.44

Inside-Plant Cables

All of the outside-plant cables are usable inside buildings except in areas where plenum or riser-rated cables are required in accordance with the National Electrical Code (NEC).

The National Electrical Code (NEC) describes minimum safety guidelines established by the National Fire Protection Association (NFPA). Article 820 describes requirements associated with Community Antenna Television and Radio Distribution Systems. Coax cables installed within buildings are tested and labeled in accordance with the NEC. State and local building code agencies should be consulted and electrical building and fire inspection organizations consulted prior to the selection, installation and operation of any cable products. All coax cables manufactured to meet the NEC requirements must be marked to identify the classification type it is listed for.

Riser-Rated Distribution Cables

Riser-rated cables feature a flame retardant polyethylene jacket. The 500 series riser-rated cable has a solid aluminum sheath. 11-series cable must be supershielded with bonded foil, 60% braid, non-bonded laminated tape and an additional 40% braid.

Despite the supershielding, 11-series cables do not have a good power carrying capacity. Therefore they should not be used in risers feeding secondary amplifiers.

Riser rated distribution cables are compared in the following.

The basic construction of the 11-series riser-rated cable is as follows.

17 gauge [Nominal 0.571 in. (1.45 mm)] copper-covered steel center conductor; foamed flame-retardant polyethylene dielectric; inner-shield aluminum-polypropylene-aluminum laminated tape with overlap bonded to dielectric; outer shield of 34 AWG bare aluminum braid wire; jacket of flame-retardant black polyvinylchloride. Nom. O.D. 0.395 in. (10.03 mm).

Physical Demensions Component	500 Series Inches	11 Series Super-Shield Inches
Nominal Center Conductor Diameter	0.109	0.057
Nominal Diameter Over Dielectric	0.45	0.280
Nominal Diameter Over Outer Conductor		
or first shield	0.500	0.287
Nominal Outer Conductor Thickness	0.025	
Nominal Diameter Over Jacket	0.566	0.405
Nominal Jacket Thickness	0.033	0.036
Maximum D.C. Resistance @ 68° f. (20° C.)		
(Inner Conductor) / 1000 ft	1.35 ohms	do not use for
(Outer Conductor) / 1000 ft.	0.37 ohms	power transfer
(Loop) / 1000 ft.	1.72 ohms	
Maximum Characteristics		
Minimum Bending Radius	6.0 in.	4.0 in.
Maximum Pulling Tension	300 lbs	130 LBS

164

The differences in attenuation are as follows:

Frequency (MHz)	500 Series Maximum dB/100'	11 Series Maximum dB/100'
5	0.16	0.36
55 (Ch. 2)	0.54	1.06
83 (Ch. 6)	0.66	1.35
187	1.01	1.98
211 (Ch. 13)	1.09	2.01
250	1.20	2.26
300	1.31	2.43
350	1.43	2.65
400	1.53	2.80
450	1.63	2.96
500	1.73	3.18
550	1.82	3.30
600	1.91	3.40
750	2.16	4.04
865	2.34	4.45
1000	2.52	4.75

Attenuation [@ 68° F. (20° C.)]

Plenum-Rated Distribution Cables

The two cable types are available in a plenum-rated version. Both the 500-series and the 11-series cable features a foamed Teflon-flourinated, ethylene propylene dielectric and a kynar jacket.

Again, supershielded construction is required when using the 11-series cable. The basic construction of the 11-series cable is changed to 14 gauge [nominal 0.064 in. (1.63 mm)] copper-covered steel center conductor; foamed Teflon dielectric (FEP); inner-shield aluminum laminated tape with overlap bonded to dielectric; outer shield of 34 AWG bare aluminum braid wire; jacket of plenum-rated material. Nom O.D. 0.351 in. (8.92 mm).

Physical Demensions Component	500 Series Inches	11 Series Super-Shield Inches
Nominal Center Conductor Diameter	0.109	0.057
Nominal Diameter Over Dielectric	0.450	0.280
Nominal Diameter Over Outer Conductor or first shield	0.500	0.287
Nominal Outer Conductor Thickness	0.025	
Nominal Diameter Over Jacket	0.524	0.372
Nominal Jacket Thickness	0.012	0.020

Maximum D.C. Resistance @ 68° f. (20° C.)
(Inner Conductor) / 1000 ft   1.32 ohms   do not use for
(Outer Conductor) / 1000 ft.   0.40 ohms   power transfer
(Loop) / 1000 ft.   1.72 ohms

Maximum Characteristics
Minimum Bending Radius   5.0 in.   4.0 in.
Maximum Pulling Tension   200 lbs   150 LBS

The attenuation of plenum distribution cables is higher since Teflon cannot be foamed to contain much air.

	500 Series	11 Series
Attenuation [@ 68° F. (20° C.)]		
Frequency	Maximum	Maximum
(MHz)	dB/100'	dB/100'
5	0.17	0.34
55 (ch.2)	0.61	1.08
83 (ch. 6)	0.78	1.30
187	1.24	2.0
211 (ch. 13)	1.35	2.18
250	1.50	2.40
350	1.85	3.0
400	2.02	3.24
450	2.16	3.51
500	2.32	3.8
550	2.46	4.0
600	2.60	4.25
750	3.43	4.9
865	3.86	5.35
1000	4.31	5.9

Service Drop Cables

Service drop cables are used between multitap ports in the riser distribution network and the outlet.

11-series, 6-series and 59-series cables can be used for service drop installations, as long as they are supershielded.
Since we covered the 11-series cable under riser distribution cables it will not be shown here.

59-series cable, however, has a very high attenuation. At 750 MHz, 100 ft. of 59-cable absorbs 7.63 dB, which is too much loss for a modern two-way HFC system. For this reason, it is not listed.

The 6-series cable is available in both riser-rated and plenum configurations. The basic construction of the riser-rated, 6-series cable is as follows:

20 gauge [nominal 0.359 in. (0.91 mm)] copper-covered, steel-center conductor; foamed flame retardant polyethylene dielectric; inner-shield aluminum-polypropylene-aluminum laminated tape with overlap bonded to dielectric; outer-shield of 34 AWG bare aluminum braid wire; jacket of flame-retardant black polyvinylchloride. Nom O.D. 0.272 in. (6.91 mm).

The basic construction of the plenum-rated 6-series is as follows:
18 gauge [nominal 0.040 in. (1.02 mm)] copper-covered, steel-center conductor; foamed Teflon dielectric (FEP); inner-shield aluminum laminated tape with overlap bonded to dielectric; outer-shield of 34 AWG bare aluminum braid wire; jacket of plenum-rated material. Nom O.D. 0.244 in. (6.20 mm).

Physical Demensions Component	6-Series Riser Rated Supershield inches	6-Series Riser Rated Supershield Inches
Nominal Center Conductor Diameter	0.035	0.032
Nominal Diameter Over Dielectric	0.180	0.143
Nominal Diameter over First Shield (Tape)	0.187	0.051
Nominal Diameter Over Jacket	0.300	0.228
Nominal Jacket Thickness	0.034	0.016

The attenuation comparison is shown in the following:

Attenuation [@ 68 F. (20° C.)] Frequency (MHz)	6- Series Riser Maximum dB/100'	6- Series Plenum Maximum dB/100'
5	0.67	0.53
55	1.86	1.58
83	2.22	1.90
187	3.18	2.88
211	3.32	3.10
250	3.58	3.40
300	3.89	3.77
350	4.15	4.15
400	4.35	4.52
450	4.60	4.86
500	4.80	5.18
550	5.00	5.45
600	5.20	5.78
750	5.80	6.60
865	6.30	7.20
1000	6.75	7.90

At 750 MHz the difference between riser rated and plenum drops is 0.8 dB. A drop length over 150 ft. will use over 10 dB of the signal provided.
If it is possible to avoid plenum conditions, do not hesitate to do so.

Broadband Amplifiers

Coaxial broadband systems consist of trunk and distribution segments. It is, therefore, not surprising that there is a differentiation made between trunk amplifiers and distribution amplifiers.

While trunk amplifiers are designed to amplify the signal over and over again, distribution amplifiers are designed for high-gain and high-output levels to feed the multitaps that supply the signals to the user.

There are numerous variations of the performance and type of amplifiers on the market. There are push-pull versions, PHD, feed forward, dual and quad output models with thermal and/or automatic level control.

The specifications listed below are offered as an example of the operational parameters.

Trunk Amplifiers

Minimum Specifications:

Pass Band, MHz	54-750	5-42	22-750	5-186

Typical Operating Conditions:

Operational Channels dB	22 or 26	10.5	22 or 26	19
Channels, Number of NTSC	110	4	53	30
Operating Levels, Recommended at Frequency, (MHz)	54-750	5-30	222-750	5-186
Input, dBmV minimum	10-10	17-17	9-9	14-14
Output, dBmV	33-27	32-32	30-27	37-37

Performance Characteristics, at Recommended Levels:
(Temperature Range -40° to +60° C)
Carrier to Interference Ratio, dB

Composite Triple Beat	84	100	82	98
Second Order Beat (F1+F2)	81	83	85	87
Cross Modulation (per NCTA standard)	94	100	94	86
Third Order Beat (F1+F2+F3)	98	110	98	108
Noise, 4MHz, 75 ohms	59.5	69	59	63
Noise Figure, dB (high channel)	9.0	9.0	11	9.0

Full Gain, dB:	25 or 28	15	25 or 28	20

Factory Alignment

Flat Loss, dB	22	15	22	20
Flatness + dB	1	0.25	0.5	0.2
Return Loss, dB minimum, all entry ports	16	18	15	16

Powering Requirements at 60 VAC, without BA:

Max AC thru currect (A)	10	10	10	10
Maximum Current (24 V dc) (Amp)	0.56	0.2	0.56	0.2

Automatic Level Control:

ALC Range, + dB at Highest Frequency	3	0	3	3

Gain Control:

Variable, dB	0.6	0-8	0-6	0-8
Plug-in Pad	0-21	0	0-21	0

All housings are cast aluminum with RF and weather seal.

Distribution Amplifiers

Distribution amplifiers are high gain amplifiers and handle output levels of +46 dBmV at the highest passband frequency.

When ordering trunk or distribution amplifiers, accessories such as pads, equalizers, load pad assemblies and interstage networks have to be ordered separately. In case your contractor is to supply amplifiers, make sure that he/she is responsible to supply such ancillary equipment.

Pads are available in steps of 0-21 dB. Equalizers are available in steps of 1.5 dB from 0, 3dB to 21 dB.

Minimum Specifications:

Pass Band, MHz	50-750	5-42	222-750	5-186

Typical Operating Conditions:

Operational Channels dB	33	15	33	20
Channels, Number of NTSC	100	4	54	30
Operating Levels, Recommended at Frequency, (MHz)	54-750	5-42	222-750	5-186
Input, dBmV minimum	13-13	17-17	13-13	17-17
Output, dBmV	46-40	32-32	46-40	37-37

Performance Characteristics, at Recommended Levels:
(Temperature Range -40° to +60° C)
Carrier to Interference Ratio, dB

Composite Triple Beat	68	91	84	81
Second Order Beat (F1±F2)	67	68	71	83
Cross Modulation (per NCTA standard)	65	76	67	78
Third Order Beat (F1±F2±F3)	75	87	81	84
Noise, 4MHz, 75 ohms	65	80.5	68.0	78
Noise Figure, dB	7.0	9.0	7.0	11.0

Full Gain, dB:	33	15	33	20

Factory Alignment, without EO:

Flat Loss, dB	32	19	36	30
Gain Slope, dB	0-2.0	0-1.5	0	0
Flatness (at gain slope)± dB	0.5	0.5	0.5	0.5
Return Loss, dB minimum, all entry ports	16	16	16	16

Powering Requirements at 60 VAC:

Power, Watts	29	28	25	28
Current (Amp)	.48	.3	0.42	0.47

Gain Control:

Variable, dB	0-8	--	0-8	0-10
Plug-in Pad	0-21	0-21	0-21	0-21

Passive Components

Splitters and Directional Couplers

Passive equipment, such as splitters, directional couplers and power inserters shall be selected on the basis of heavy construction, corrosion resistance, DC power bypass and following electrical characteristics.

Minimimum Specifications of Splitters and Directional Couplers

Pass Band	5-750 MHz
Pass Band Flatness	± -.25 dB
Impedence (all ports)	75 ohms
Cable Connectors	Center Seizure: Housing tapped 5/8"; 24 TPI for standard entry port connectors. Center seizures accommodate up to 0.172" diameter center conductor.
Cable Power Bypass	12 A MAX
Hum Modulation	Better than -70 dB at 10 amperes current

Specifications:

Frequency Range:	5 MHz to 750 MHz
Frequency Response:	Cable equivalent, all ports +0.25 dB
Return Loss:	5 MHz to 10 MHz, 18 dB 10 MHz to 450 MHz, 20 dB 450 MHz to 750 MHz, 64 dB
Hum Modulation at 10 A:	5 MHz to 500 MHz, 70 dB 500 MHz to 550 MHz, 67 dB 550 MHz to 750 MHz, 64 dB
Power Passing:	10 A, 60 volts, 60 Hz

Typical performance of splitters and directional couplers is listed below.
Splitters come in 2-way and 3-way configurations. Directional couplers are
available for uneven tap losses of 8, 12, 16, and 20 dB.

750 MHz	Splitters	and		Directional Couplers		
Insertion Loss	2-way	3-way	DC-8	DC-12	DC-16	DC-20
5-9MHz	3.70	3.8/7.5	1.40	0.90	0.70	0.50
10-200	3.70	3.8/7.1	1.40	0.90	0.70	0.50
201-300	3.80	4.0/7.2	1.50	1.00	0.80	0.50
301-400	3.90	4.3/7.4	1.60	1.10	0.80	0.60
401-500	4.00	4.6/7.6	1.80	1.30	1.10	0.70
501-600	4.30	4.9/7.9	2.10	1.60	1.60	0.80
601-750	4.50	5.1/8.1	2.30	1.80	1.60	1.00
Tap Loss						
5-9 MHz	3.7+.5	7.5+.5	8.8+.5	12.0+.5	16.0+.5	20.0+.5
10-600	4.0+.5	7.6+.5	8.5+.5	12.0+.5	16.0+.5	20.0+.5
601-750	4.3+.5	8.0+.5	9.0+.5	12.3+.5	16.3+.5	20.4+.5
Isolation						
5-9 MHz	25	24/25	20	22	24	70
10-450	30	28/25	23	27	30	67
451-500	27	28/25	23	27	30	65
501-550	27	28/25	22	27	30	63
551-600	27	27/25	22	24	28	61
601-750	25	24/25	20	22	23	50
Return Loss						
5-9 MHz	18	17	18	17	17	18
10-450	20	20	20	20	20	20
451-500	18	18	18	18	18	18
501-550	18	18	18	18	18	18
551-750	17	17	17	17	17	18
Current Capacity	10 A	10 A	10 A	10 A	10 A	10 A
Hum Mod.	70 dB	70 dB	70 dB	70 dB	70 dB	70 dB

Power Inserter Specifications:

Frequency Range:	5 MHz to 750 MHz
Frequency Response:	Cable equivalent, all ports + 0.25 dB
Return Loss all Ports:	5 MHz to 10 MHz, 18 dB
	10 MHz to 450 MHz, 20 dB
	450 MHz to 750 MHz, 64 dB
Hum Modulation at 10 A:	5 MHz to 500 MHz, 70 dB
	500 MHz to 550 MHz, 67 dB
	550 MHz to 750 MHz, 64 dB
Power Passing:	15 A, 60 V ac max input port
	10 A, 60 V ac max output port

Multitaps:

Two, four and eight multitaps may be used at building-entrance and at IDF locations for service drop cable connections. Multitaps are 2-way devices and permit transmisssion and reception of the frequency band.

Minimum Specifications:

Frequency Range:	5 to 750 MHz	
Frequency Response, Cable Equivalent, All Port	± 0.35 dB	
Return Loss all Ports:	5-400 MHz	20 dB
	400-500 MHz	18 dB
	500-750 MHz	17 dB
Hum Modulation at 6 A:	5-450 MHz	70 dB
	450-500 MHz	68 dB
	500-750 MHz	64 dB
Power Passing:	6 A, 60 Volts, 60 Hz	
Tap to Tap Minimum Isolation:	5-500 MHz	25 dB
	500-750 MHz	23 dB

Since 2-port multitaps are not commonly used in the densities of HFC systems, only 4 and 8-port multitaps are shown with their respective tap and insertion losses.

4-port Multitaps

Tap Loss	MHz	8	11	14	17	20	23	26	29	32	35
Tolerance ± dB			1	1	1	1	1	1	1	1	1
Insert Loss	5		3.3	1.8	0.9	0.7	0.6	0.5	0.4	0.4	0.4
	50		2.8	1.4	0.7	0.5	0.4	0.4	0.3	0.3	0.3
	100		3.1	1.4	0.7	0.5	0.4	0.4	0.3	0.3	0.3
	300		3.1	1.5	0.8	0.7	0.6	0.5	0.4	0.4	0.4
	400		3.3	1.7	1.0	0.8	0.7	0.6	0.5	0.5	0.5
	500		3.6	2.1	1.4	1.1	0.8	0.8	0.7	0.7	0.7
	600		3.8	2.3	1.6	1.3	1.1	1.0	0.8	0.8	0.8
	700		4.1	2.6	1.8	1.6	1.3	1.1	1.0	1.0	1.0
	750		4.3	2.8	2.0	1.8	1.5	1.3	1.2	1.3	1.3
Isol. Tap toOut	5-599	24	24	27	30	33	36	38	40	42	44
	600-750	22	22	25	28	31	34	36	38	40	42
Isol. Tap toTap	5-49	24	24	24	24	24	24	24	24	24	24
	50-599	26	26	26	26	26	26	26	26	26	26
	600-750	24	24	24	24	24	24	24	24	24	24
Ret. Loss in/out	5-49	17	17	17	17	17	17	17	17	17	17
	50-599	18	18	18	18	18	18	18	18	18	18
	600-750	16	16	16	16	16	16	16	16	16	16
Ret. Loss Tap	5-49	17	17	17	17	17	17	17	17	17	17

	50-599	18	18	18	18	18	18	18	18	18	18
	600-750	16	16	16	16	16	16	16	16	16	16
Hum Mod 6Amp	5-599	-70	-70	-70	-70	-70	-70	-70	-70	-70	-70
	600-750	-64	-64	-64	-64	-64	-64	-64	-64	-64	-64

8-port Multitaps

Tap Loss	MHz	12	15	18	21	24	27	30	33	36	39
Tolerance ±dB		1	1	1	1	1	1	1	1	1	1
Insert Loss	5		3.3	1.8	0.9	0.7	0.6	0.5	0.4	0.4	0.4
	50		2.8	1.4	0.7	0.5	0.4	0.4	0.3	0.3	0.3
	100		3.1	1.4	0.7	0.5	0.4	0.4	0.3	0.3	0.3
	300		3.1	1.5	0.8	0.7	0.6	0.5	0.4	0.4	0.4
	400		3.3	1.7	1.0	0.8	0.7	0.6	0.5	0.5	0.5
	500		3.6	2.1	1.4	1.1	0.8	0.8	0.7	0.7	0.7
	600		3.8	2.3	1.6	1.3	1.1	1.1	0.9	0.9	0.9
	700		4.1	2.6	1.8	1.6	1.4	1.2	1.1	1.1	1.1
	750		4.3	2.8	2.0	1.8	1.6	1.4	1.3	1.3	1.3
Isolation Tap to out	5-599		27	30	33	36	38	40	42	44	46
	600-750		25	28	31	34	36	38	40	42	44
Isolation Tap to Tap	5-49	24	24	24	24	24	24	24	24	24	24
	50-599	26	26	26	26	26	26	26	26	26	26
	600-750	24	24	24	24	24	24	24	24	24	24

Ret. Loss in/out	5-49	17	17	17	17	17	17	17	17	17	17
	50-599	18	18	18	18	18	18	18	18	18	18
	600-750	16	16	16	16	16	16	16	16	16	16
Ret. Loss Tap	5-49	17	17	17	17	17	17	17	17	17	17
	50-599	18	18	18	18	18	18	18	18	18	18
	600-750	16	16	16	16	16	16	16	16	16	16
Hum Mod. 6Amp	5-599	-70	-70	-70	-70	-70	-70	-70	-70	-70	-70
	600-750	-64	-64	-64	-64	-64	-64	-64	-64	-64	-64

Coaxial-Cable Connectors

The selection of high-quality connectors is an issue of importance. Connectors are used to interconnect the coaxial distribution cable with the equipment housings. While all housing ports are standardized, the make-up of the various cables require different sizing of sleeves, center conductor seizing and locking of the outer sheath.

The connector, especially when not installed correctly, is the main source of ingress and egress problems. The FCC has set limits for signal radiation from coaxial systems which must be observed. Since our system is designed for an upper frequency of 750 MHz, only proven and radiation-free connectors can be used.

Housing Connectors

Chassis mounting connectors are designed for standard 5/8 - 24 equipment entry. The connector must be a pin-type connector with internal seizing.

The connector must have an integral sleeve in the nut assembly to seize the outer conductor and fully shield the seized connection.

The pin has a diameter of 0.067 in. and can be 1.6 in. or 2.31 in. long. The dimensions of the three hex nut assemblies are 1 3/8 in. front and center and 1 1/4 in. for the backnut depending on the cable type.

Housing-to-Housing Connectors

Any interconnection of two devices shall be made using a housing-to-housing connector. The housing-to-housing connector is designed to eliminate the need for coaxial jumpers. It features a pin of typical 1.6 in. in length on both sides to permit the seizing of the center conductor in each device.

Housing-to-housing connectors permit the cascading of multitaps, amplifiers and passive devices. To keep the devices in the same orientation, the use of non-rotational housing-to-housing connectors is recommended.

The dimensions are 0.75 in. for both the center and the front hex nut.

Housing Terminations

Housing terminations are used to terminate the last device into a 75 ohm termination. At the same time the housing terminator has a blocking capacitor to block the 60 Vac power from burning out the termination resistor.

The mechanical make-up is identical to the housing connector. Both center and outer conductor seizing is provided as in the pin type connector.

Service Drop Connectors

The F-connector is the commonly used connector system for service drops. Both ends of the 6-series drop cables require male F-connectors for the connection to the tap ports as well as to the wall-plate outlet assembly.

The new universal F-connector, type F-6-AHS/USA, permits the connectorization of different size drop cables. When used with a 0.5 in. sleeve, the connector will provide a radiation-free connection. A 0.360 hex crimp tool is required to complete the connection.

F-Terminations

F-Terminations are F-male connectors with a built-in 75 ohm terminating resistor. Every open port of a multitap must be terminated. AC-blocking types are available, if required.

Power Supplies for Coaxial Cables

Standard 60 VAC, 15 or 20 Amp CATV-type power supplies shall be used to accommodate the powering of active devices in the Cable TV and ITV systems through the cable. The power supply translates 110 VAC to the 60 VAC on the coaxial cable.

The power supply unit shall be wall board-mounted and interconnected with the coaxial cable through a power-inserter. JCA cable shall be used between the power supply output port and the power-inserter unit. Cable power supplies connect to standard 110 Vac power outlets.

HFC Access Equipment

RF Modulators

Modulators are used at the outputs of a matrix switch, a computer terminal or at a camera location to convert baseband video and audio to RF channels on the HFC distribution system. Modulators shall be frequency agile from 50 to 750 MHz. Frequency agility is required to accommodate change of channel allocation that may be required in the future. If fully automated operation is required, the RF modulators shall contain an RS232C port for operation from a computer terminal.

The modulators shall be a high-quality CATV headend design, accommodating adjacent channel transmission and feature a 1 RU profile.

Typical specifications require a maximum output level of +60 dBmV with spurious outputs of < -60 dBmV, an adjustable sound carrier to -25 dB below video carrier and a vestigal sideband response of -20 dB at channel edge. The video input level shall be 0.5 V p-p for 87.5% modulation.

Suppliers of modulators are numerous. The recommended model should be HRC/IRC switchable, stereo-compatible and in a 1 RU profile for all frequencies.

Minimum Specifications

RF:

Frequency Range	Channels 2 to 94
	50 to 750 MHz (HRC, IRC, Standard)
Frequency Accuracy	± 5 KhZ
Output Level	60 dBmV minimum
Spurious	-60 dBc maximum
Output Impedance	75 ohms

Output Return Loss	14 dB min within the output channel	
Sound Carrier Level	Adjustable from -5 to -25 dB relative to picture carrier	

C/N Ratio	Typical	Minimum
In-Band	70 dB	67 dB
Adjacent Channel	74 dB	70 dB
Wideband	78 dB	75 dB

IF:

Picture IF Output Frequency	45.75 MHz
Picture IF Output Level	35 dBmV nominal
Sound IF Output Frequency	41.25 MHz
Picture IF OUtput Level	10 to 30 dBmV
CW IF Output Frequency	45.75 MHz
CF IF Output Level	5 dBmV $\pm$ 5.0 dB

GENERAL:

Power Requirements	30-35 Watts
Weight	12.8 lbs
Dimensions	19"W x 1.75"H x 17"

VIDEO:

Standard Baseband	0.5 to 2.0 V p-p for 87.5% modulation
Input Level Range:	
Encoded Video Input Level	1.75 V p-p nominal for 87.5%
Video Input Impedance	75 ohms
Video Input Return Loss	30 dB minimum
K-Factor	2% maximum wirh 2T pulse
S/N ratio (Lum Weight)	64 dB at 87.5% mod.
Chroma Delay, relative to Standard	
Precorrection:	
15° to 25° C	$\pm$ 50 nSec
0° to 50° C	$\pm$ 65 nSec
Frequency Response	$\pm$ 0.5 dB maximum from 25 Hz to 4.1 MHz
Differential Gain	$\pm$ 0.25 at 87.5% mod.
Differential Phase	$\pm$ 0.5 degrees maximum
Hum and Noise	-60 dB at 87.5% mod.
Tilt	1.0% maximum per NTC-7, 3.3

AUDIO:

Input Level Range	Dual Input range, selectable
Range 1:	-10 to + 10 dBm
Range 2:	+5 to +25 dBm
Input Impedance	Dual Input Range Selectable
Low Impedance	600 ohms balanced
High Impedance	Greater than 10 kohms
Frequence Response	$\pm$ 1.0 dB maximum from 30 Hz to 15 KHz
Harmonic Distortion	.0% max. from 30 Hz to KHz at $\pm$25 KHz dev.
FM Hum and Noise	-60 dB max. with respect to $\pm$ 25 KHz dev.
Intercarrier Frequency	4.5 MHz $\pm$ 500 Hz

RF Demodulators

RF demodulators are used to demodulate off-air, cable television or HFC channels to baseband video and audio.

To be able to assign the demodulator to any cable TV channel, the demodulator shall be frequency agile. To enable remote control from the computer terminal, the demodulator shall have a hard-wired remote control connection for RS-232C computer operation.

The demodulator shall be of CATV headend quality, use synchronous/envelope detection and feature low-noise figure, a minimum of 60 dB adjacent channel and image rejection and feature adjustable video to 1.5V p-p and audio to 2 RMS 600 ohm balanced outputs.

Suppliers of demodulators are numerous. Models equipped with an RS-232C port and computer control are available.

Minimum Specifications

RFInput:

Input Impedance	75 ohms
Input Return Loss	16 dB minimum
Input Level Range	-20 to +30 dBmV
Noise Figure Full Gain	5 dB max. sub and VHF low channel
	6 dB max. mid-high, super band VHF anf 750 MHz

Channel	
Carrier to Noise Ratio at	60 dB minimum - VHF
+ 10 dBmV input	57 dB minimum - UHF
Image Rejection	60 dB minimum - VHF
	40 dB minimum - UHF

Intermodulation at Input	-80 dB or better for equal level adjacent channels at + 110 dBmV
Cross Modulation at Input	-80 dB or better for equal level adjacent channels at +20 dBmV
IF Output Level	+ 30 dBmV
AGC Regulation	+ 0.5 dB maximum change for input variation from -20 to +30 dBmV

VIDEO:

Output Type Rear Panel	Two 75 ohm type "F" connectors
Output Return Loss	30 dB minimum all outputs
Output Level at 87.5%	1 volt p-p: adjustable to 2 volts p-p minimum Depth of Modulation
Amplitude Response	± 0.5 dB maximum 30 Hz to 4 MHz
Differential Gain	± 2% maximum, 87.5% Depth of Modulation, 10% to 90% APL
Differential Phase	± 0.5 degrees maximum, 87.5% Depth of Modulation, 10% to 90% APL

12.5 T Pulse Chroma Delay	$\pm$ 50 nanoseconds
Field Square Wave Tilt	1% maximum
Video Signal-to-Noise Ratio	56 dB VHF
+ 10 dBmV RF Input Level	53 dB UHF
Residential 4.5 MHz	-60 dB maximum
non-combined Operation	

AUDIO:

Output Type - rear panel	600 ohm balanced type "XLR" connector and screw terminals
Output Type - front panel	600 ohm single ended screw terminal, low impedance tip jacks
Output Level	Adjustable +6 dBm maximum $\pm$25 KHz Deviation
Distortion	1% maximum 50 Hz-15,000 Hz $\pm$ 25 KHz deviation +6 dBmV output maximum
Flatness	+ 1 dB, 50 Hz-15,000 Hz including 75 usec de-emphasis

4.5 MHz Aural:

Outputs	a) Separate 75 ohm type "F" connector
	b) Combined with 2md video output, switch provided for addition or removal of 4.5 MHz from video output No. 2 as required
Output Level	0.1 volt p-p nominal adjustable to 0.5 volt p-p maximum combined and separated outputs

General:

| AC Power Requirements | 100 to 130 V, 50 to 60 Hz, 25 W |
| Operating Temperature Range | 32° F to 120° F (0° C to 48.9° C) |

Matrix Switching Equipment

The size of the matrix routing switch is directly related to the required number of video inputs or sources and the number of HFC distribution channels desired. The matrix routing system shall be non-blocking and permit the selection of any output or any input. All switching of video with audio follow-on shall be in the vertical interval.

An initial complement with 16 inputs and 16 outputs is recommended. The system shall be expandable by the addition of routing-switch shelves to 32 and 48, or more, source inputs and to 32 and 48 or more outputs.

The control of all switching functions of the matrix routing system shall be RS-232C from the computer keyboard for real-time and time-programmable operation.

The video-audio matrix switch shall feature a 60 MHz bandwidth and follow the electrical specifications outlined below. The baseband transmission through the swtich must support analog and digitial services.

180

Minimum Specifications

VIDEO:

Input Impedance	75 ohms loop-through
Frequency Response	± 0.1 dB 100 KHz to 5 MHz, ± 0.5 dB to 60 MHz
Bandwidth	3 dB to 60 MHz typical
Gain	Unity, ± 0.05 dB
Crosstalk	< -63 dB at 10 MHz (all inputs hostile)
Output DC Offset	Blanking at OV ± 50 mv
Differential Gain	< 0.2% at 1V p-p output, 10-90% APL
Differential Phase	< 0.2 deg. at 1V p-p output, 10-80% APL
Hum and Noise	< -70 dB
Input Return Loss	> 40 dB to 10 MHz
Output Return Loss	< 40 dB to 10 MHz
Tilt	< 0.1% at line or field rate

AUDIO:

Input Impedance	> 20K ohms, balanced
Output Impedance	600 ohms, balanced
Frequency Response	± 0.5 dB 20 Hz to 30 KHz
Bandwidth	3 dB to 100 KHz
Gain	Unity
Crosstalk	< 86 dB to 20 KHz at + 30 dBm
Stereo Separation	> 90 dB
Hum and Noise	< -80 dBm
Total Harmonic Distortion	< 0.05% 20 Hz to 30 KHz
Input, Output Level	+ 27 dBm max.
Connectors	Pin Header, 0.1" centers

A 16 x 16 video matrix switch shall require a rack space of 1 RU (2 RU for video and audio - monaural or stereo). An expanded 32 x 32 video/audio switch shall not occupy more than 8 RU.

Chapter 9

The HFC Broadband Network Design Process - Inside-Plant Design -

As communications manager of an enterprise, you certainly have been exposed to very adverse conditions at times. When the telephone system required upgrading, the cutover scheduling and timing had to be worked out to the finest details.

When every department requested the installation of LANs, competition between the various interest groups gave you sleepless nights. Designing a high-speed FDDI data network to interconnect these islands of automation may be the next highest priority on your agenda.

But there are also a number of people that are calling for enterprisewide desktop videoconferencing. Does this mean that the data network requires upgrading to SONET rings with Gbit/s speeds and ATM switches at node locations? Not necessarily. The answer may be the establishment of an HFC broadband network that can carry multiple bi-directional, high-quality, full-motion video NTSC or MPEG-2 formats and carry high-speed data at DS-3 or T-I speeds. An HFC system can also give you the ability to carry 24 channels of voice in a T-I format between departments or PBX switching equipment.

Even if you are not an engineer, or your expertise is administration or conceptual planning, you can design an HFC broadband system. As long as you can add, subtract and multiply, you can design your own broadband network. This design guide takes you by your hand and leads you step by step through each category and calculation.

An HFC system, as the name hybrid fiber-optic/coaxial explains, consists of two separate and different topologies:

a)   The Broadband Coaxial System
b)   The Fiber-optic Transmission System

The broadband coaxial system is always located at the end of an HFC system. Coaxial cable is best used to provide service to many users.

The fiber-optic transmission system is always at the beginning of an HFC system. Fiber is best used to overcome long distances.

Since the design of the coaxial broadband system is more involved and since it has more topological variances, we will concentrate on the broadband coaxial design phase first.

Inside-Plant Design - From Service Drops to the Building Entry Location

The design process of the in-building broadband coaxial system includes all service drops and the riser distribution cable and equipment.

Since every building entry location is the starting point of the in-building network, it is recommended to locate either a fiber-optic receiver or a distribution amplifier at the entry location. This amplifier will isolate the inside plant from the outside plant. Inside-plant design can now proceed without any consideration for outside-plant particulars.

The advantages of the separation of inside- and outside-plant design areas are numerous:

-        the inside plant can be modified and extended without affecting the outside-plant network

184

-   the outside plant can be modified, rearranged and extended
    without affecting any in-building network

-   changing the outside plant from coaxial to fiber service to feed
    a building does not affect in any way the in-building design

-   the outside-plant network can grow to serve new buildings
    without any affect on initial service locations

The broadband network design process described in this chapter is a step-by-step design guide and assures a high-quality broadband network that will serve bi-directional, multichannel voice, data and video services for many decades to come.

Designing the Service Drop

Outlet Levels

The outlet level is a signal value that is expressed in dBmV or decibels based on one millivolt. dBmV is used exclusively as the unit of measure for all broadband measurements at RF frequencies.

The RF frequency range of a broadband coaxial network cable extends from O to IOOO MHz. The electronic equipment that amplifies the RF signal restricts this bandwidth to a range of 50 to 750 MHz in the forward direction.

When providing bi-directional transmission bandwidth, the industry settled on two common but different standards:

a)   the Sub-low Transmission System
     with a forward bandwidth of 50 to 750 MHz
     and a reverse bandwidth of 5 to 42 MHz

b)   the High-split Transmission System
     with a forward bandwidth of 222 to 750 MHz
     and a reverse bandwidth of 5 to I86 MHz

Based on these standards, we must look at the outlet levels at different operating frequencies. We also must allow for tolerances in our level determination, since splitter and coupling losses can not be precisely assured. The outlet level also has to meet or exceed FCC requirements, which has determined the level to a TV set and shall be at least O dBmV.

To accommodate one additional splitter at the outlet, an additional 4 dB is required, and to permit return transmission, a 55 dBmV level must be accommodated.

As a result, the recommended outlet level specification is as follows and must be met for any outlet in the system:

For forward systems between 50 to 750 MHz: +10.0 ±4.0 dBmV
For forward systems between 222 to 750 MHz: +8.0 ±3.0 dBmV
For return transmission from 5 to 186 MHz: +55.0 ±1.0 dBmV

Cable Selection

The listing of qualifying drop cables in Chapter 8 covers available cable types suitable for utilization as service drops. There are major categories of service drop cables:

a)   Riser-rated cable
     which is standardized under NEC Article 820 - type CATVR

b)   Plenum-rated cable
     which is standardized under NEC Article 820 - type CATVP

In the design checklist for building No. 158, you have already segregated the two conditions. In fact, it is very common to experience plenum conditions in a segment of a building. The remainder drop wiring must be riser rated to comply with prevailing fire regulations. The CATVR specification requires the jacket of the cable to consist of flame retardant PVC (polyvinyl-chloride). While plenum-rated cable can be used as a substitute for riser-rated cable, plenum open air return areas require the use of only CATVP-type cables.

There are three different sizes of service drop cables available:

a)   RG-59 type drop cable
     with a diameter of 0.24 inches

b)   RG-6 type drop cable
     with a diameter of 0.3 inches

c)   RG-11 type drop cable
     with a diameter of 0.405 inches

The Shielding Properties

The diameters listed above refer to super-shielded cables, which consist of a bonded foil, followed by a braid, a nonbonded laminated tape and an additional

186

braid. The major difference between CATVR riser-rated and CATVP plenum-rated cables is the dielectric and the jacket. In the case of the riser-rated cable, a polyvinylchloride (PVC) jacket is used, while the plenum cable features a foamed teflon dielectric and a plenum-rated jacket.
Table 9-I shows the differences in the mechanical makeup.

Cable Type	Center Gauge	First Braid %	Second Braid %	Jacket
CATVR-59	22	53	35	PVC
CATVR-6	20	40	60	PVC
CATVR-11	17	60	40	PVC
CATVP-59	20	50	40	Teflon
CATVP-6	18	60	60	Teflon
CATVP-11	14	60	60	Teflon

Table 9-I - Super-shielded Drop Cables

Any one of these cables has the desired shielding properties to hold ingress and egress to acceptable levels. Ingress of a local broadcast transmitter can produce co-channel interference, which are horizontal lines caused by the different path lengths of the received signal. Local high-powered FM stations can produce beat products in TV pictures.

The most important reason, however, for using super-shielded drop cables in the HFC system is egress. A modulator, connected to a transmit outlet, transmits at a level of +55 dBmV. A part of this energy would radiate into the air if standard shielded drop cable were to be used. The FAA and FCC require strict compliance with the radiation limits of coaxial cables since many cable frequencies are also used for air-to-air and air-to-ground communication systems. For this reason, the use of super-shielded drop cables is considered mandatory in two-way transmission systems and highly recommended for forward only HFC systems operating at a design frequency of 750 MHz.

The Attenuation Properties

In the selection process of a suitable drop cable, it is important to remember that

a)   the smaller the cable diameter,
      the higher the attenuation

b)   the higher the transmitted frequency,
      the higher the attenuation

So, before making a cable selection, it is important to analyze the attenuation differences of the various cables.

Table 9-2 shows typical attenuation figures for operating frequencies at 5, 50, 180, 220 and 750 MHz. These frequencies have been chosen because they are representing the design frequencies for forward and two-way sub-split and high-split transmission systems.

Attenuation per 100 ft	5 MHz (dB)	50 MHz (dB)	180 MHz (dB)	220 MHz (dB)	750 MHz (dB)
CATVR-59	0.89	2.41	4.08	4.40	7.63
CATVR-6	0.67	1.86	3.18	3.40	5.80
CATVR-11	0.36	1.06	1.98	2.15	4.04
CATVP-59	0.65	1.95	3.55	3.80	7.90
CATVP-6	0.53	1.58	2.88	3.20	6.60
CATVP-11	0.34	1.08	2.00	2.25	4.90

Table 9-2 - Attenuation per 100 ft. at various frequencies

A review of this table shows that plenum-rated cables have lower attenuation at low frequencies and higher attenuation at high frequencies than riser-rated cables. The reason, of course, is the teflon dielectric used in all plenum-rated cables.

The table also shows that the attenuation of an ll-type cable is about one-half of a 59-type cable.

The Handling Properties and Costs

The first reaction then is to select ll-type cables for the service drops. There are, however, a number of negative factors that need to be considered:

-    high costs
-    handling difficulties

The ll-type cable, especially the plenum-rated variety, is extremely hard to work with. To terminate an ll-type drop in a shallow outlet box and make the connection to the F-type faceplate connector is not an easy task.

188

Since both the 59 and the 6-type cables are quite flexible and can be handled with relative ease, the decision is now down to price. Obviously, the 59-type cable is less expensive. But let us take another look at the attenuation at 750 MHz. The attenuation of the 59-type cable is about 1.5 dB greater than the 6-type cable. This means that the level required in the riser must be higher to accommodate the higher loss service drops. This means that more amplification is required. So, by opting for the less expensive 59-type cable, we may have to add an amplifier here and there in the in-building riser network and the cost advantage is negated.

Taking all the factors into consideration, i.e. attenuation, handling ease and costs, the best compromise appears to be the riser-rated CATVR-6 and, if needed, the plenum-rated CATVP-6 cables.

The 150 ft. Limitation

From the attenuation table 9-2 we can determine the attenuation of the service drop at different lengths.

For instance, a 100 ft. CATVR-6 drop has an attenuation of 5.8 dB at 750 MHz and of 0.67 dB at 5 MHz.

Conversely, a 150 ft. CATVR-6 drop has an attenuation of 1.5 x 5.8 dB or 8.7 dB at 750 MHz., and 1.5 x 0.67 dB or 1.0 dB at 5 MHz. If we look at the attenuation difference, we find that the 100 ft. drop has a difference of 5.13 dB and the 150 ft. drop has a difference of 7.7 dB or 2.57 dB higher.

Let us look at a 250 ft. long drop. The attenuation at 750 MHz is 2.5 x 5.8 dB or 14.5 dB. The attenuation at 5 MHz is 2.5 x 0.67 dB or 1.675 dB. The difference results to over 12.8 dB.

This type of attenuation difference between highest and lowest transmission frequencies results in uncontrollable outlet levels. A ±6 dB level swing between high and low frequencies cannot be tolerated as it may affect picture quality and prevent the control of return-transmission levels.

The desirable outlet levels have been previously defined to be

+10.0 ±4 dBmV for 50 to 750 MHz

+8.0 ±3 dBmV for 220 to 750 MHz

+55.0 ±1.0 dBmV for 5 to 180 MHz return transmission

These outlet level specifications cannot be met using 250 ft. long service drops. The selection of 150 ft. for a service drop standard is a compromise. The system will still work if you standardize on 200 ft. service drops, but it will

189

be more difficult to control forward and return-transmission levels.

Standardization to 150 ft. means that all service drops are cut to this length. Even in a case where only a 30 ft. drop is required, the full length of 150 ft. is connected and the unused portion is coiled.

The standardization of the service drop length reduces design complexities and eliminates wide level variations at outlets as well as at the headend. This is not to say that there may be exceptions, i.e. a single 180 or 200 ft. special drop. Such exceptions, however, must be calculated individually and require separate test specifications for levels at different frequencies. Because of all the additional problems that are encountered when special service drop lengths are required, it is recommended to use the 150 ft. service drop length as a must standard.

In cases where the shape of the building and the location of the vertical riser are such that longer footages are required, it is recommended to extend the riser horizontally to meet the 150 ft. service drop requirement.

Forward Level Calculations

The 150 ft. drop length will greatly simplify the design of the service drop. Only one calculation need to be made. This calculation can then be applied to all multitap levels throughout the coaxial system.

The forward design should be made for all forward frequencies that will be encountered:

a)   Forward System 50 to 750 MHz (Riser Rated)

Outlet Level Specification +10.0 ±4.0 dBmV		
Frequency	50 MHz	750 MHz
Attenuation of 150 ft.		
CATVR-6 drop	2.8 dB	8.7 dB
Level required		
at begin of drop	+12.8 dBmV	+18.7 dBmV
Tolerance	± 4.0 dB	± 4.0 dB

The calculation shows that the outlet level specification of +8.0 ±3.0 dBmV can be met by providing an input level to the 150 ft. drop of +8.8 to +16.8 dBmV at 50 MHz and from +14.7 to 22.7 dBmV at 750 MHz

b) **Forward System 50 to 750 MHz (Plenum Rated)**

Outlet Level Specification +10.0 ±4.0 dBmV

Frequency	50 MHz	750 MHz
Attenuation of 150 ft.		
CATVP-6 drop	2.37 dB	9.9 dB
Level required		
at begin of drop	+12.37 dBmV	+19.9 dBmV
Tolerance	± 4.0 dB	± 4.0 dB

The calculation shows that the outlet level specification of +10.0 ±4.0 dBmV can be met by providing an input level to the 150 ft. drop from +8.37 to +16.37 dBmV at 50 MHz and from +15.9 to 23.9 dBmV at 750 MHz

c) **Forward System 220 to 750 MHz (Riser Rated)**

Outlet Level Specification +8.0 ±3.0 dBmV

Frequency	220 MHz	750 MHz
Attenuation of 150 ft.		
CATVR-6 drop	5.1 dB	8.7 dB
Level required		
at begin of drop	+13.1 dBmV	+16.7 dBmV
Tolerance	± 3.0 dB	± 3.0 dB

The calculation shows that the outlet level specification of +8.0 ±3.0 dBmV can be met by providing an input level to the 150 ft. drop from +10.1 to +16.1 dBmV at 220 MHz and from +13.7 to +19.7 dBmV at 750 MHz

d) **Forward System 220 to 750 MHz (Plenum Riser)**

Outlet Level Specification +8.0 ±3.0 dBmV

Frequency	220 MHz	750 MHz
Attenuation of 150 ft.		
CATVP-6 drop	4.8 dB	9.9 dB
Level required		
at input of drop	+12.8 dBmV	+17.9 dBmV
Tolerance	± 3.0 dB	± 3.0 dB

The calculation shows that the outlet level specification of +8.0 ±3.0 dBmV can be met by providing an input level to the 150 ft. drop from +9.8 to +15.8 dBmV at 220 MHz and from +14.9 to +20.9 dBmV at 750 MHz

Forward Calculation Summary

The forward level calculations for the 150 ft. service drop are summarized in the following.

To simplify the design process, an average max./min. level is shown, which can be used for RR or plenum drops.

For an outlet specification of +10.0 ±4.0 dBmV, the input level of the 150 ft. service drop must meet the following max./min. requirements and a recommended average max./min. level:

	50 MHz		750 MHz	
	from	to	from	to
		(dBmV)		(dBmV)
CATVR-6	8.8	16.8	14.7	22.7
CATVP-6	8.37	16.37	15.9	23.9
Recommended Average	8.8	16.4	14.7	23.9

For an outlet specification of +8.0 ±3.0 dBmV, the input level of the 150 ft. service drop must meet the following max./min. requirements:

	220 MHz		750 MHz	
	from	to	from	to
		(dBmV)		(dBmV)
CATVR-6	10.1	16.1	13.7	19.7
CATVP-6	9.8	15.8	14.9	20.9
Recommended Average	9.8	15.8	13.7	20.9

It is useful to memorize these recommended averaged max./min. level values. All of them will be used over and over again when calculating multitap values in the riser network. The recommended average max./min. level at the input of the service drop is at the same time the port level of any multitap in the system. At the same time, the recommended average max./min. level provides us with the ground rules for system equalization. Looking at these numbers, we can already see that the riser distribution system requires higher levels at higher frequencies and can only barely tolerate flat levels. System designs that allow lower frequency levels to be higher than high-frequency levels will make it impossible to meet recommended outlet specifications.

Return-Level Calculations

The sub-low return in a 50 to 750 MHz forward system uses the 5 to 42 MHz band of frequencies. The return transmission in a high-split system occurs between 5 and 186 MHz.

The return-level calculation of the 150 ft. service drop then concerns itself with only three frequencies, i.e.

- the highest return frequency
  in the high-split system (180 MHz)

- the highest return frequency
  in the sub-low system (50 MHz)

- the lowest return frequency
  of both the sub-low and the high-split system (5 MHz)

The return-level calculation assumes that a transmission device is connected to the transmit outlet port.

The transmission device is typically a modulator that modulates a 1 V p. to p. video signal to the desired RF channel. Modulators have an output level range between +50.0 and +60.0 dBmV. The commonly used output level setting is +55.0 dBmV, which is the starting point of our calculation.

The return-level calculation determines the level at the riser end of the 150 ft. drop, i.e. the signal level that is provided to the multitap port at 5, 50 and 180 MHz.

Using the cable attenuation figures from Table II-2, we can determine the following losses for the 150 ft. drop. In addition to the attenuation of the cable, we also have to include a 4 dB loss for the splitter in the outlet that separates transmit from receive ports.

a)   Return-Transmission Calculation of 150 ft. Drop
      (Riser Rated)

Modulator Output Level	+55.0±2.0dBmV		
Splitter Loss	-4.0 dB		
Level entering the			
150 ft drop	+51.0±2.0dBmV		
Frequencies	5 MHz	50 MHz	180 MHz
Attenuation			
using CATVR-6 cable	1.0dB	2.8dB	4.77dB
Resulting Level			
at Multitap port{dBmv}	+50.0	+48.2	+46.2
Tolerance	±2.0	±2.0	±2.0

We can see that the return-transmission level varies with the frequency of transmission. The higher the transmit frequency, the greater the attenuation. The resulting return-transmission levels, after passing through the 150 ft. drop, are in the following ranges:

- at 5 MHz from +48.0 to 52.0 dBmV
- at 46 or 50 MHz from +46.2 to 50.2 dBmV
- at 180 MHz from +44.23 to 48.23 dBmV

b)   Return-Transmission Calculation of 150 ft. Drop
      (Plenum Rated)

Modulator Output Level			+55.0 dBmV ±2.0 dBmV
Splitter Loss in Outlet		-4.0 dB	
Level entering the 150 ft. Drop		+51.0 dBmV ±2.0 dBmV	

Frequencies	5 MHz	50 MHz	180 MHz
Attenuation of 150 ft. Drop			
using CATVP-6 cable	0.8 dB	2.37 dB	4.32 dB
Resulting Level			
at Multitap port	+50.2	+48.63	+46.68
Tolerance	±2.0 dB	±2.0 dB	±2.0 dB

The return transmission levels, again, vary with frequencies but are quite similar to the riser-rated cable. The resulting return-transmission levels, after passing through the 150 ft. drop, are in the following ranges:

- at 5 MHz from +48.2 to 52.2 dBmV
- at 46 or 50 MHz from +46.63 to 50.63 dBmV
- at 180 MHz from +44.68 to 48.68 dBmV

Return-Level Summary

Since it is evident that only minor differences exist between riser-rated and plenum-rated drop cables, we can simplify our design calculations by, again, developing an average recommended port level for return transmissions. All return-transmission levels, after passing through a 150 ft. service drop, will enter the multitap port at the following average levels:

- at the lowest frequency (5 MHz) at +50.0 ±2.0 dBmV
- at the highest sub-low frequency (46 MHz) at 48.0 ±2.0 dBmV
- at the highest high-split frequency (180 MHz) at 46.0 ±2.0 dBmV

The calculation of the return-transmission levels are important when considering coaxial and fiber-optic return-transmission equipment along the path to the headend. As we will see in the riser design, the input level of all return signals requires a level of +17.0 dBmV when entering the return amplifier. As we can see from the above, the level variation between the high frequency of the high-split system (I80 MHz) and the lowest frequency (5 MHz) can only be controlled within 4 dB. From +46.0 to +50.0 dBmV $\pm$2.0 dBmV. Therefore, the +I7.0 dBmV input level to the reverse-transmission amplifier must be given a tolerance of at least $\pm$3.0 dBmV.

The design of the service-drop section of the HFC broadband network is now complete and we can focus our attention on the design of the riser distribution network.

Designing the Riser Distribution Network

The Building-Entry Location
or the Single-Building Headend

The building entry location accepts the outside plant network and acts as the headend and control location for the in-building portion of the network.

In a single-building network, there is no building-entry location for your private HFC network, but there must be a headend for the control and operation of the in-building broadband network. In such a case, the headend also represents the gateway to the outside world.

The building entry location, in a broader sense, then can be the multipurpose location that controls the in-building broadband network. It can assume any one of the following functions:

- the interface between the outside and inside world
  in a campus HFC network
- the headend for the in-building network
- the control center of the network in a single building
- the gateway to the outside world in a single building
- the switching center for in-building desktop videoconferencing
- the editing studio and authoring center in a single building

Even though we like to think that the building-entry location is somewhere in the basement, there is no reason that it could not be located elsewhere in the building. The important consideration that must be made in the selection of a suitable location is that it represents the heart of the in-building network and that all broadband cabling originates at that location.

Two-way broadband voice, data and video, for instance, translate their frequencies at the headend. A transmission received from any connected location is up-converted at the headend to a downstream frequency assignment.

In the case of our hypothetical campus, of course, the building-entry location does not require any equipment other than the building-entry amplifier. The headend or video operations center is located elsewhere in Building No. 132. All voice, data or video translations take place at that location.

Our hypothetical scenario of Fig. 7-2 shows the riser plan of Building No. 158 and assumes that the building-entry location is room B-14. It is this riser network that we want to design first.

Cable Selection

As in the case of the service drop cable, we must first determine which cable types are suitable and practical.

Since riser cables are used to transport the signals to the service drops, we are looking for a cable that has no attenuation so that the service drop levels are all the same. This, of course, is not possible, but it illustrates the desire to use a low-attenuation cable to better overcome the distance requirements.

The listing of cables in Chapter 8 covers available cable types suitable for in-building riser networks.

There are, again, two major categories of riser cables:

    a)   Riser-rated cable, as it has been standardized under NEC Article 820, type CATVR

    b)   Plenum-rated cable also standardized under NEC Article 820, type CATVP

There are two different sizes of riser cables available:

    a)   The RG-11 type cable with a diameter of 0.405 inches

    b)   The 500 type cable with a diameter of 0.566 inches

We have dealt with the RG-11 type cable already under the service drop category. At that time, we dismissed it for use as a service drop.

Can CATVR-11 or CATVP-11 be used in a Riser Network?

The answer to that question is a qualified yes. It can be used where attenuation is not a problem and the distances are short. It cannot be used when secondary amplifiers have to be placed in the riser. The CATVR-11 or CATVP-11 cables have a high DC resistance and do not carry 60 Vac on the center conductor well.

The 500-type cables utilize a solid outer conductor sheath and a 0.109 inch center conductor, which reduces the DC resistance to 1.35 ohms/1000 ft. and provides for 100% shielding.

Table 9-3 shows the mechanical properties of the two cable types.

Cable Type	Center Conductor	Shielding	Jacket
CATVR- 11	17 ga	2 braids	PVC
CATVP- 11	14 ga	2 braids	Teflon
500 JCAR	0.109"	solid	PVC
500 JCAP	0.109"	solid	Teflon

Table 9-3 - Riser Cable Types

The Attenuation Properties

Obviously, the 11-type cable is more flexible and less expensive than the 500-type cables, but when it comes to attenuation, they are quite superior. Especially when system design calls for an upper frequency of 750 MHz, the extra expense may be a wise investment and may reduce the number of amplifiers.

Table 9-4 shows typical attenuation figures for 100 ft. of cable for the operating frequencies of 5, 50, 180, 220 and 750 MHz.

Attenuation per 100ft (dB)	5 MHz	50 MHz	180 MHz	220 MHz	750 MHz
CATVR- 11	0.36	1.06	1.98	2.15	4.04
CATVP- 11	0.34	1.08	2.0	2.25	4.9
500 JCAR	0.16	0.52	1.0	1.11	2.16
500 JCAP	0.17	0.58	1.24	1.38	3.43

Table 9-4 - Riser Cable Attenuation per 100 ft.

A review of this table shows that the riser-rated 500-type cable has almost one-half the attenuation of the 11-type cable. This means that twice the distance

197

can be overcome. The 5OO-type plenum version, however, shows a steep increase of attenuation at high frequencies. This attenuation increase is caused by the foamed teflon dielectric required for plenum-rated cable. It is, therefore, recommended to use the 5OO-type plenum cable for short sections of horizontal risers and to change over to RR-rated cable whenever possible.

The Selection of Amplifiers

It has been determined before that the building-entry location or the begin location of the in-building riser distribution network requires amplification.

Section 8 provides for typical specifications for coaxial distribution amplifiers as well as for fiber-optic receivers. Since we are concerned first with Building No. I58 in our hypothetical campus system, we know that the building is fed by coaxial cable.

If Building No. I58 were a fiber-node location, we would require a fiber-optic receiver followed by a coaxial distribution amplifier.

What are the important considerations in the amplifier selection process?

We are looking for an outdoor type, high-gain block with a good output level. Ideally, the amplifier features a forward bandwidth of 46 to 75O MHz for sub-low operation and a bandwidth of 22O to 75O MHz for high-split operation. The unit should have a typical gain of 32 to 36 dB and provide for a forward output level of +46 dBmV.

There are also dual amplifiers available that can be used to feed additional feeder cables. These amplifiers are seldom required for in-building riser topologies.

A forward output level of +46 dBmV minus the gain of the amplifier requires us to provide an input level of about +I4 dBmV. This value is also important for the outside-plant design process, which will be discussed at a later point.

In the return direction, the amplifier needs to be equipped with a reverse amplifier module either for operation between 5 to 42 MHz (sub-split system) or between 5 to I86 MHz (high-split system). The gain of the 5 to 42 MHz reverse module is typically about I5 dB so that for an output level of 32 dBmV all the reverse transmission inputs are to arrive at a +I7.O dBmV level. In the case of the high-split reverse module, operating between 5 and I86 MHz, the output level is usually around +37.O dBmV with a gain of 2O.O dB. So, again, all reverse transmissions have to arrive at the reverse amplifier with a level of +I7.O dBmV.

Table 9-5 lists the major electrical requirements of both forward and return distribution amplifiers.

Amplifier Specs.(dB)	46-750 MHz	220-750 MHz	5-50 MHz	5-186 MHz
Frequency Response	±0.75	±0.75	±0.35	±0.35
Min. Full Gain	33	33	15	20
Noise Figure	7.0	7.0	9.0	11.0
Output Level	46	46	32	37
Tilt	9	9	flat	flat
Cross Mod.	60.0	60.0	100.0	86.0
Triple Beat	56	56	---	92

Table 9-5 - Distribution Amplifier Requirements

The Selection of Passives and Multitaps

We looked at cable types and their attenuation over frequency and we looked at the properties of typical amplifiers. The third group of HFC components are the passives.

Passive devices consist of splitters, directional couplers and multitaps:

- A splitter divides a signal into two or three separate paths. The splitting loss is equally divided between the two or three branches.

- A directional coupler also divides a signal into two separate paths, but the splitting loss between the two branches is unequal.

- A multitap provides 2, 4 or 8 ports for the connection of service drops. In a way, the multitap is also a directional coupler. It reduces the through-going signal somewhat and provides various values of attenuation to a group of ports.

The Mechanical Properties

When selecting passive components, it is recommended to stay away from indoor and apartment-type devices. Both the mechanical and electrical properties of these devices do not comply with the standards that must be applied in building a broadband HFC network for 75O MHz operation.

Section 8 covers the technical data of all passive devices. When looking for longevity, it is important to consider components of proven performance for many years in the outdoor environment. Passive devices have undergone a number of evolutionary changes to arrive at the present level of development. Some of the important mechanical requirements are listed below:

- the casing must be a durable aluminum casting, consisting of the housing and the faceplate

- a polymer coating is required for environmental protection

- the faceplate must be removable without having to disconnect the housing from the cable

- housing and faceplate must have a tongue and grove mechanical and electrical protection to be weatherproof and radiationproof

- the housing connectors must have a center conductor seizure of l5 in.lb. to 20 in.lb.

- the housing closure torque must be better than 50 in.lb.

- the housing design must permit aerial strand mounting, pedestal mounting and mounting on wallboards.

The Electrical Properties

Even more important in the selection of passive components are the electrical requirements to assure the quality of performance.

The first requirement, of course, is the passband. All passives must operate to the stated specifications between 5 and 750 MHz.

The second most important requirement is the tap-loss accuracy. There are many passives available that will not guarantee the stated tap loss within ±l dB. You cannot design a broadband network without trusting the published tap-loss values.

The third requirement is the insertion loss. The lower the insertion loss, the higher your signal level and the farther you can take the signal. So, before purchasing any passives, secure the insertion-loss tables of the various vendors of passive equipment, compare the insertion losses over the frequency band and pick the one supplier that features the lowest insertion loss at any frequency and the lowest rise of insertion loss at 750 MHz.

There are other important considerations that cannot be overlooked in the selection process of passive equipment:

- the return loss of all ports of a splitter or directional coupler shall be better than 18 dB over the frequency range. A high return loss prevents reflections of the signal in the return direction.

- the tap-port to tap-port isolation must be better than 23 dB over the frequency range. The higher the isolation value between ports, the less interaction between TV sets connected to the outlets.

There have been instances where system designers used T-connectors to daisy-chain broadband outlets. A T-connector has no isolation between ports at all and the interaction between TV sets becomes noticeable when switching from one channel to another.

In addition, the power-passing capability of the passive device is of importance. An additional amplifier in the riser network must be powered through the coaxial cable. A minimum of 6 Amp at 60 Vac power-passing capability is recommended for both splitters and multitaps.

Table 9-6 shows the relationship of tap losses and insertion losses over the frequency range of splitters and directional couplers in decibels.

Passives	Tap Loss	Insertion Loss 5MHz	Insertion Loss 50 MHz	Insertion Loss 180 MHz	Insertion Loss 220 MHz	Insertion Loss 750 MHz
Splitter	2-way	3.8	3.7	3.75	3.8	4.9
Splitter	3-way	6.3	5.7	5.75	5.8	7.2
Splitter	3-way high/low	3.8/6.3	3.7/5.7	4.0/6.0	4.1/6.1	4.9/8.2
Dir.Coupl	8	1.4	1.5	1.55	1.6	2.4
Dir.Coupl	12	0.7	0.9	0.95	1.1	2.0
Dir.Coupl	16	0.6	0.7	0.75	0.9	1.7

Table 9-6 - Insertion Losses of Splitters and Directional Couplers

We said before that multitaps come in 2-port, 4-port and 8-port versions. Since the cost of a 2-port multitap is almost identical and the insertion loss is almost the same, the 2-port tap can safely be eliminated from consideration. Even if your initial drop count at a particular floor is one, there is always the possibility of future expansion. As a matter of fact, it may be good practice to determine the total number of possible outlets on a floor and to design the riser for ultimate expansion.

Table 9-7 shows the relationship of tap losses and insertion losses of 4-port and 8-port multitaps over the frequency range.

Tap-loss values may vary slightly from one manufacturer to the other. For instance, a 23-value tap with 8 ports may be a 24-value tap with another manufacturer. Generally, it is possible to substitute one value for another. But before doing so, assure yourself about the ±l.O dB tap-loss variation specification.

Four Way Taps	Tap Loss or Value	5 MHz	50 MHz	180 MHz	220 MHz	750 MHz
	32	0.3	0.5	0.5	0.5	1.2
	29	0.3	0.5	0.5	0.5	1.2
	26	0.4	0.5	0.6	0.6	1.3
	23	0.4	0.5	0.6	0.6	1.3
	20	0.5	0.7	0.8	0.8	1.6
	17	0.9	0.9	1.1	1.1	1.8
	14	1.8	1.6	1.8	1.8	2.8
	11	3.2	3.0	3.2	3.2	4.3
	8	Terminate	Terminate	Terminate	Terminate	Terminate
Eight Way Taps						
	32	0.3	0.5	0.5	0.5	1.2
	29	0.4	0.5	0.6	0.6	1.3
	26	0.5	0.7	0.8	0.8	1.6
	23	0.5	0.7	0.8	0.8	1.6
	20	1.0	1.0	1.2	1.2	1.8
	17	1.8	1.8	1.8	1.8	2.6
	14	3.2	3.0	3.5	3.5	4.1
	11	Terminate	Terminate	Terminate	Terminate	Terminate

Table 9-7 - Insertion Losses and Tap Values of Multitaps

The table shows increased insertion losses at higher frequencies. Passives and multitaps then act similar to cable and can tilt the transmission levels to favor the low end. The conclusion that must be reached from the review of the table is that outlet specifications require level tolerances. Too many multitaps installed in series will distort the system performance and not permit the meeting of any outlet level specification since the low frequencies will be transmitted at higher levels.

Many broadband cable systems have been designed using the high-frequency values only as a basis for the design. In order to assure high-quality system performance of a broadband HFC network, the design task must include both the highest and lowest passband frequencies in both directions.

The Riser Design Process for Transmission in the Forward Direction

The Design of the Building No. 158 Riser Network

In order to get organized and to be able to design a system segment efficiently, it is recommended to make a preliminary sketch of the design particulars.

Fig. 9-I shows the riser route of Building No. 158. Originally, the drop lengths in the 5th floor were compiled on Fig. 7-I. Fig. 7-2 shows the riser routing within the building. Fig. 9-I combines the data with the components that we need to provide service to the 31 service drops.

In the building-entry location, we have placed an amplifier with test taps at both input and output. To use low-attenuation cable, we have selected 500 JCAR riser-rated 500-type cable. Since we are serving 5 drops in the first floor, we have assigned an 8-port tap. Another 20 ft. of 500-type cable gets us to the 2nd floor where we need another 8-port tap for seven locations. In the 3rd floor, a 4-port tap will suffice for 2 drops. In the 4th floor, another 4-port tap is required for 2 drops. When we go to the fifth floor, we need a splitter or directional coupler for the plenum-rated horizontal riser to 2 multitaps. The service drops of room 519 and one drop of room 518 are combined in an 8-port tap. The remaining 4 drops are connected to the last 4-port tap. The splitter's second port feeds a 4-port tap for fifth-floor service drops and another 4-port tap in the sixth floor.

What information do we need for this design?

First of all, the forward design must identify the type of system that we are designing. Is it a 50 to 750 MHz sub-low system or are we designing a 220 to 750 MHz high-split system? By concentrating on the three frequencies 50, 220 and 750 MHz, we can design for both cases.

All we need now are the amplifier output levels, the cable attenuation numbers and the passive device data of the foregoing pages.

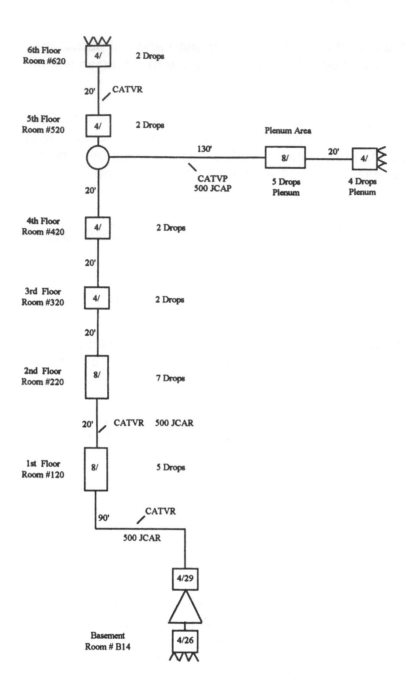

Fig. 9-1 Riser Route - Building 158

The design calculations should always proceed from the amplifier outbound to the extremities of the system.

Step I - Amplifier Setup

Table 9-5 shows the amplifier electrical data. The output level is shown as +46 dBmV. We know that low frequencies are not attenuated as much as high frequencies. We can tilt the amplifier output to +46 dBmV at 750 MHz to +41 dBmV at 220 MHz and to +40 dBmV at 50 MHz.

So, our starting levels are:

	50 MHz	220 MHz	750 MHz
Amp. Output	+40.0	+41.0	+46.0

Step 2 - The Test Tap

There is a test tap 4/29 in our drawing of Fig. 9-I. The 4/29 tap has the following insertion losses as indicated in Table II-7:

	50 MHz	220 MHz	750 MHz
4/29 Tap	-0.5	-0.5	-1.2

By subtracting the tap insertion losses from the amplifier output levels, we are determining the signal level into cable:

| Level into cable | 39.5 | 40.5 | 44.8 |

Step 3 - The First 90 ft. Cable Section

Table 9-4 shows us the attenuation of the 500-JCAR cable for a 100 ft. length. Since we have only 90 ft., we multiply the data with a factor of 0.9.

	50 MHz	220 MHz	750 MHz
First 90 ft.	-0.47	-1.0	-1.94

By subtracting these cable losses from the level into the cable, we obtain the signal levels at the input to the first floor multitap:

	50 MHz	220 MHz	750 MHz
Level to first floor multitap	39.03	39.5	42.86

Step 4 - Selection of the First Floor Multitap

When we determined the level requirement for the 150 ft. service drop, we arrived at level ranges for each frequency. It is now required to recall this information:

	50 MHz	220 MHz	750 MHz
Desired level into 150 ft. drop	9 to 16.2	10.5 to 15.8	15.0 to 19.5

By subtracting the desired average signal level for the 150 ft. drop from the level to the first floor multitap, we get numbers ranging from 31 to 23. Our selection of a tap must meet the level range of the service drop at all frequencies. The selection of a 26-value, 8-port tap seems to fit the requirement. We subtract 26 from the arriving cable level and obtain the port level of the tap.

	50 MHz	220 MHz	750 MHz
Port levels at 8/26 tap	13.03	13.5	16.68

This is the signal level that will meet both the +10.0 ±4.0 dBmV and the +8.0 ±3.0 dBmV outlet level specifications.

Before we can move to the second floor, we have to calculate the insertion loss on the riser cable. The value 26 tap has the following insertion losses:

	50 MHz	220 MHz	750 MHz
Insertion loss of 8/26 tap	-0.7	-0.8	-1.6

By subtracting these losses from the Step 3, first floor multitap level, we can now determine what signal we are sending to the 2nd floor.

	50 MHz	220 MHz	750 MHz
Level into cable to 2nd floor	38.33	38.7	41.26

Step 5 - The Second Cable Section

Again, we have to deduct the loss created by the 20 ft. of cable to determine the signal levels for the 2nd floor multitap (multiply Table 9-4 figures by .2):

	50 MHz	220 MHz	750 MHz
20 ft. cable	-0.104	-0.222	-0.432

By subtraction, we arrive at 2nd floor cable levels:

	50 MHz	220 MHz	750 MHz
Level to 2nd floor multitap	38.226	38.478	40.828

Step 6 - The Second Floor Tap

Can we use a 26-value tap again? It appears that we cannot since the drop level would be below 15 dBmV. So, the choice will be an 8/23 tap. By subtracting 23 from the cable level, we obtain the port level:

	50 MHz	220 MHz	750 MHz
Port level 8/23 tap 2nd floor	+15.226	+15.478	+17.828

This signal level meets the criteria, but already shows that the tilt between 220 and 750 MHz has reduced itself from 5 dB to 2.35 dB and we are not at the 3rd floor yet. But let us continue to the 3rd floor. First, we need the insertion loss of the 8/23 tap and then the 20 ft. cable section. Let us put them both together:

	50 MHz	220 MHz	750 MHz
Insertion loss of			
8/23 tap	-0.7	-0.8	-1.6
20 ft. cable	-0.104	-0.222	-0.432

If we subtract both numbers from the 2nd floor level numbers of Step 5, we obtain 3rd floor levels:

	50 MHz	220 MHz	750 MHz
Level to 3rd floor multitap	37.422	37.456	38.796

Step 7 - The Third Floor Tap

At the 3rd floor, we are using a 4-port tap. The value of this tap can, again, be a 23, resulting in a port level of

	50 MHz	220 MHz	750 MHz
Port level 4/23 tap, 3rd floor	14.422	14.456	15.796

The port levels are still within the desired range, but considering that we have another 230 ft. to go and have to encounter the insertion loss of 3 more taps in series, it is already very apparent that we will not be able to meet outlet specifications in the 5th floor.

What has been done wrong and how can we remedy the situation without having to insert another amplifier?

To find the answer, we have to first find out how far we can go at 750 MHz and still be within the desirable service drop level range. The question is: how much cable and how many taps use up our 5 dB tilt between 220 and 750 MHz?

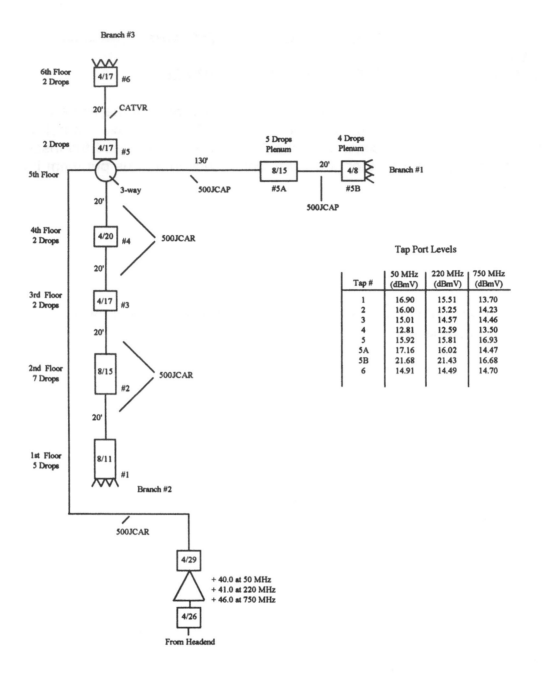

Fig. 9-2 Final Riser Design - Building 158

Let us assume we are using 300 ft. of 500-JCAR cable. The attenuation at 750 MHz is 2.16 dB per 100 ft. and the attenuation at 220 MHz is 1.11 dB per 100 ft. The difference then is 1.05 dB per 100 ft.; 300 ft. of cable gives us a tilt difference of 3.15 dB.

This, by itself is not bad, but let us assume that we need four 8-port multitaps somewhere in this 300 ft. distance. Looking at the tap insertion losses in Table 9-7, we can see that between 220 and 750 MHz approximately .8 dB are required per tap. For the four taps, this amounts to 3.2 dB. Adding the two together, we have a tilt requirement of 6.35 dB. Our amplifier tilt of 5 dB

208

compensates for most of this roll-off, but we have to realize that we can go no further without re-equalizing the system by using a cable equalizer and an additional amplifier.

Is there a more cost-effective solution then using an additional amplifier?

In deed, there is. Fig. 9-2 shows the replacement design for Building No. 158.

From the amplifier at B 14, we have taken an uninterrupted riser cable to the fifth floor. There, we use a splitter and back-feed to the first floor with one branch and serve the fifth and sixth floor plant from the other branch.

One branch now consists of the uninterrupted riser of 170 ft., plus four 20 ft. sections, for a total of 250 ft. and it has 4 taps.

The other branch consists of the same 170 ft. long riser, plus 150 ft. of plenum cable, for a total of 320 ft., but with only 2 taps and 1 splitter.

Since we are close to the limits of 300 ft. and 4 taps, we will have to redesign the Building No. 158.

Redesign of the Building No. 158 Riser Network

Redesign -
Step 1 - Amplifier Setup and Riser to the Fifth Floor

We use the same amplifier alignment procedure:

	50 MHz	220 MHz	750 MHz
Amp. Output	+40.0	+41.0	+46.0
4/29 Test tap	-0.5	-0.5	-1.2
170 ft. of JCAR	-0.884	-1.887	-3.672
3-way Splitter at 5th floor	-5.7	-5.8	-7.2
Level for Branches 1, 2 and 3	+32.916	+32.813	+33.928

Redesign -
Step 2 - The Longest Run (Branch I)

If we can make the horizontal riser in the fifth floor meet the service drop ranges, we shall not have any problems with the other two branches.

	50MHz	220MHz	750MHz
130 ft. of 500 JCAP	-0.754	-1.794	-4.459
8/15 Value Tap Insertion	-3.0	-3.5	-4.1
20 ft. of 500 JCAP	-0.116	-0.276	-0.686
	29.046	27.243	24.683
4/8 Value Tap	8.0	8.0	8.0

The resulting tap port levels are:

for the 8/15 Tap	17.162	16.019	14.469
for the 4/8 Tap	21.683	21.243	16.683

We did the best we could. At the 8/15 tap, the 750 MHz level is a little low and does not meet the specification for plenum drop cable by 0.431 dB. At the terminating 4/8 tap, the 750 MHz level is within specification, but the 220 and 50 MHz levels are about 5 dB too high. But overall, a good compromise has been reached.

Outlet levels that have been designed out of spec. should be recorded so that the implementation contractor is not blamed for your design compromises.

Redesign -
Step 3 - The Vertical Riser (Branch 2)

Using the levels from our Step I redesign, we can now determine the tap values for floors 4, 3, 2 and I:

	50 MHz	220 MHz	750 MHz
Level 5th floor for branches 1,2,3	+32.916	+32.813	+33.928
20 ft. 500-JCAR	-0.104	-0.222	-0.432
Level at tap input	+32.812	+32.591	+33.496
4/20 port levels	+12.812	+12.591	+13.496*
4/20 Insertion	-0.7	-0.8	-1.6
20 ft. 500-JCAR	-0.104	-0.222	-0.432
Level at tap input	+32.008	+31.569	+31.464
4/17 port levels	+15.008	+14.569	+14.464**
4/17 Insertion	-0.9	-1.1	-1.8
20 ft. 500-JCAR	-0.104	-0.222	-0.432
Level at tap input	+31.004	+30.247	+29.232
8/15 port levels	+16.004	+15.247	+14.232**
8/15 Insertion	-3.0	-3.5	-4.1
20 ft. 500-JCAR	-0.104	-0.222	-0.432
Level at tap input	+27.9	+26.525	+24.7
8/11 port levels	+16.9	+15.525	+13.7**

All 50 and 220 MHz service drop levels are within the recommended ranges.
The 750 MHz level is low by 0.24 dB for a 220 to 750 MHz design*.
The other 750 MHz levels are low by 0.236 and 0.468 dB**.
Since the measurement accuracy of the best signal level meter is within I dB,
we can be quite pleased with the results.

Redesign -
Step 4 - The Sixth Floor Riser (Branch 3)

This calculation should not present any problematic conditions. We start again
with the level for branches of Step I:

	50 MHz	220 MHz	750 MHz
Level			
for branches 1,2,3	+32.916	+32.813	+33.928
4/17 port levels	+15.916	+15.813	+16.928
4/17 Insertion	-0.9	-1.1	-1.8
20 ft. 500-JCAR	-0.104	-0.222	-0.432
Level at tap input	+31.912	+31.491	+31.696
4/17 port levels	+14.912	+14.491	+14.696

This completes the design of the Building No. 158 riser distribution system. It is
good practice to record the design data on the riser design diagram, as this has
been done in Fig. 9-2. The sketch summarizes all pertinent information as to

> Number of drops
> Location of drops
> Type of cable
> Length of cable
> Splitters
> Multitap - sizes and numbering
> Multitap values
> Port levels of each tap for 50, 220 and 750 MHz

It is, of course, recommended that you keep the design calculations too so that
you can refer to them when the technician comes in to show you what he
measured.

The Design of a High-Rise Building

In the riser design for Building No. 158, we determined that due to higher
attenuation at higher frequencies the low-frequency signal increases with
distance and curtails meeting the outlet specifications. We also observed that
each 100 ft. of 500-JCAR cable and each device, whether multitap or splitter,
have about a 0.8 dB higher loss at 750 MHz than at 220 MHz.

These observations led to the conclusion that 300 ft. of cable and four taps in series is all that we can expect to feed from one amplifier. We put this knowledge to good use by center-feeding three branches with a combined footage of 340 ft. and a total of 8 multitaps.

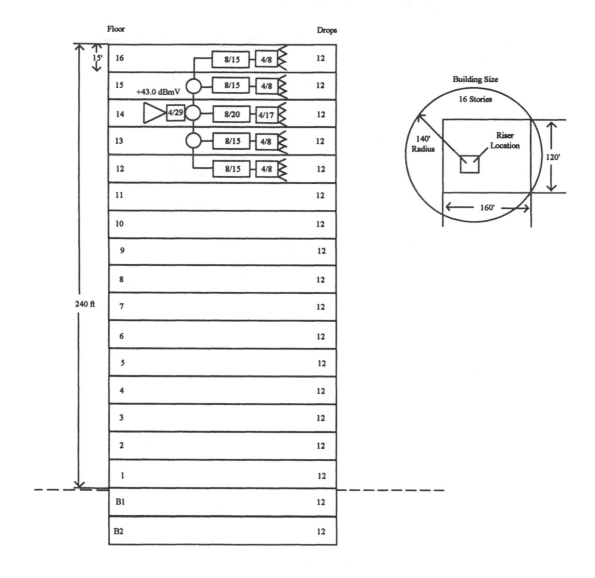

Fig. 9-3 High Rise Building Layout - with Sample Design

Can symmetrical center-fed riser architecture minimize the number of amplifiers?

The answer is yes. Feeding the center of a group of multitaps allows us to branch into as many directions as we have to simply by using splitters or directional couplers to form the required number of branches.

If we go one step further and add symmetry to each of the branches, we can optimize level utilization, maximize the number of outlets served and offer a highly cost-effective solution for the design of a riser distribution network.

Fig. 9-3 shows the makeup of a hypothetical high-rise building. The plane view shows the dimensions of the building and the location of the riser. By drawing a 140 ft. circle around the riser location, we can determine that we can reach all proposed outlet locations with 150 ft. long service drops. The building has 16 floors and there are 12 outlets per floor and in the basement Bl. The total number of outlets then is 204.

We know that with 12 drops per floor, we need as a minimum two multitaps, one 8-port and one 4-port.

How many floors can we serve with one amplifier?

Starting from the top, the sixteenth floor, we have two devices. All other floors will have three devices, because we need a splitter. Each splitter uses up 4.9 dB. Our amplifier can only be set for +43.0 dBmV at 750 MHz, because there are at least two amplifiers in cascade. The three-way splitter next to the amplifier has 7.2 dB of attenuation.

The Trial Design

As a trial design, we place the amplifier at the fourteenth floor and feed two floors up and down. Obviously, the extreme locations, i.e. the sixteenth and twelfth floor are worst cases. Starting with the amplifier output level, we can check the workability of the design architecture by calculating only one floor. Symmetry helps to reduce the design time as well.

	50 MHz	220 MHz	750 MHz
Amplifier output	+37.0	+38.0	+43.0
4/29 Test tap	-0.5	-0.5	-1.2
3-way Splitter	-5.7	-5.8	-7.2
15 ft. 500-JCAR	-0.08	-0.17	-0.32
2-way Splitter	-3.7	-3.8	-4.9
15 ft. 500-JCAR	-0.08	-0.17	-0.32
Level at 8/15 tap	+26.96	+27.56	+29.06
Port level	+11.96	+15.56	+14.06

Insertion	-3.0	-3.5	-4.1
Level at 4/8 tap	+23.96	+24.06	+24.96
Port level	+15.96	+16.06	+16.96

This design seems to work out quite well. All the service drop input-level specifications are met. We could even change the cable to CATVR-II cable, which would reduce the tap-port level by 0.64 dB at 750 MHz.

Are all the tap values identical on all floors?

At the fourteenth floor, where the amplifier is located, we are using only one 3-way splitter. So our level is about 5 dB higher, which should qualify for the selection of a higher tap value.
Let us confirm that

	50 MHz	220 MHz	750 MHz
Amplifier output	+37.0	+38.0	+43.0
4/29 Test tap	-0.5	-0.5	-1.2
3-way Splitter	-5.7	-5.8	-7.2
Level at 8/20 tap	+30.8	+31.7	+34.6
Port level	+10.8	+11.7	+14.6
Insertion	-1.2	-1.2	-1.8
Level at 4/17 tap	+29.8	+30.5	+32.8
Port level	+12.8	+13.5	+15.8

The result of this trial design shows that one amplifier can take care of five floors, 10 multitaps and 60 outlets within the specification requirements. Whether this design is used for a 50 to 750 MHz sub-split system or for a 220 to 750 MHz high-split system, all outlet specifications will be met.

Service to other Floors

The trial design covered floors 12 to 16 only. The same equipment configuration is required for floors 7 and 11, and for floors 2 and 6. Since the equipment layout features center-feeding and symmetry, it can be used over and over again.

Outlets in basement B1 and in the first floor can be served from the amplifier at the building-entry location.

The only missing link is the riser cable that feeds the three amplifiers on floors 4, 9 and 14. This is easily accomplished by a parallel feeding riser. We simply take the output of the building-entry amplifier and after using a splitter to feed B1 and floor 1, we bring it up to floor 9, then split and feed floors 4 and 14. All three amplifiers at floors 4, 9 and 14 are in parallel. This means, they are not in series or in cascade. All amplifier feeds are perfectly symmetrical.

Fig. 9-4 shows the complete riser diagram with all multitap values. There are never more than two amplifiers in cascade. One of the primary responsibility of the designer is to avoid the cascading of amplifiers. A parallel feed riser system goes a long way towards this goal.

The Symmetrical Riser Design

First of all, we have to select the cable. Since we need to send 60 Vac through the cable to power the amplifiers, there is only one choice - the 500-JCAR cable.

Second, we have to determine the exact footages between the building-entry location and the other amplifiers. Fig. 9-4 shows the cable-routing diagram with distances and devices.

The design calculation is shown in the following:

	50 MHz	220 MHz	750 MHz
Ampl. at building entry	+37.0	+38.0	+43.0
4/29 Test tap	-0.5	-0.5	-1.2
3-way Splitter	-5.7	-5.8	-7.2
200 ft. 500-JCAR	-1.04	-2.22	-4.32
3-way Splitter	-5.7	-5.8	-7.2
Level to 9th floor Ampl.	+24.06	+23.68	+23.08
75 ft. 500-JCAR	-0.39	-0.83	-1.62
Level to 4th floor Ampl.	+23.67	+22.85	+21.46

The same level applies to the 14th floor amplifier.

Are these the correct input levels for amplifiers?

No, they are not. In Table 9-5, we recorded amplifier requirements. We also know that we have to derate the output level to +43.0 dBmV because we have 2 amplifiers in cascade. The amplifier has a gain of 33 dB. If we subtract this from the output, we should deliver +13.0 dBmV to the amplifier.

The amplifier at the 9th floor requires a 10 dB pad at its input. The amplifier at the 4th and 14th floor each need 9 dB pads.

Pads are plug-in types and have to be ordered separately from amplifiers. A good selection, during system activation, taxes the capabilities of the implementor. Pads are available in 1dB steps.

We also need one other plug-in item - the equalizer. We can see from the input levels that there are only minor differences between high and low frequencies. This means that the amplifier input levels are flat. However, we are looking for a 6 dB higher output level at the highest frequency. This preemphasis of the

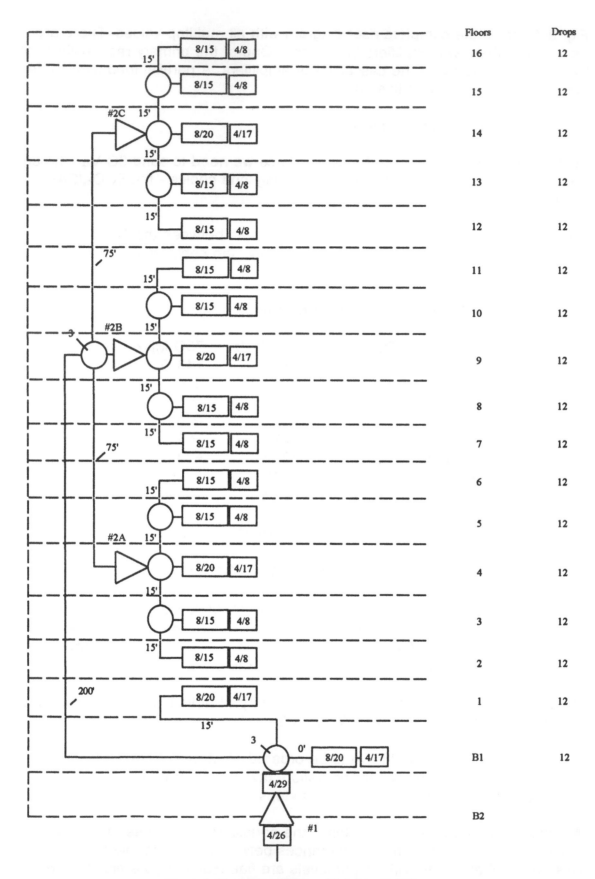

Floors	Drops
16	12
15	12
14	12
13	12
12	12
11	12
10	12
9	12
8	12
7	12
6	12
5	12
4	12
3	12
2	12
1	12
B1	12
B2	

Fig. 9-4 Final Riser Design - High Rise Building

higher frequency is provided by equalizers that simply attenuate the lower frequencies along a linear curve. The effect is called "tilt" or "slope".

Again, it is wise to carry a good selection of equalizers when attempting to align amplifiers and set operational levels. In our example, the equalizer values of 6 to 9 dB should produce the desired preemphasis of the 750 MHz level. Equalizers are plug-in and available in 1.5 dB steps.

The Forward Transmission Design of a Large Horizontal Building

Large horizontal buildings are not at all different from high-rise buildings except that the riser cabling is in the horizontal plane. The same groundrules apply for the design:

> -determine were the outlets are and group them into clusters of 40 to 60 outlets

> -decide on an amplifier location to feed each cluster in the center and as symmetrical as possible

> -interlink the cluster amplifiers with distribution cable and feed the amplifiers in a parallel configuration to avoid more than two amplifiers in cascade

> -calculate the amplifier input levels and determine pad and equalizer values

Fig. 9-5 shows a hypothetical riser design for a large horizontal one-story building with only 2 amplifiers in cascade. When installing such a system, it is recommended to make all amplifier locations accessible for maintenance. While multitaps can be located in false-ceiling spaces, the amplifiers should not. Wallboard-mounting of amplifiers and associated passive devices is recommended. Such wallboards may be located in closets or in hallway sections where they can be protected by suitable enclosures.

Fig. 9-5 shows a #1 building-entry amplifier and 12 secondary amplifiers. All secondary amplifiers are in a parallel configuration and numbered by riser-number identification. There are three risers with 4 amplifiers each. The amplifier identification number shows first that the amplifier is the second in cascade (2), then the riser identification (A, B or C), and then the number for all amplifiers in a riser (1, 2, 3 or 4).

The layout of risers A, B and C has been arranged in a totally symmetrical manner. For each riser the configuration is identical and consists of a 3-way

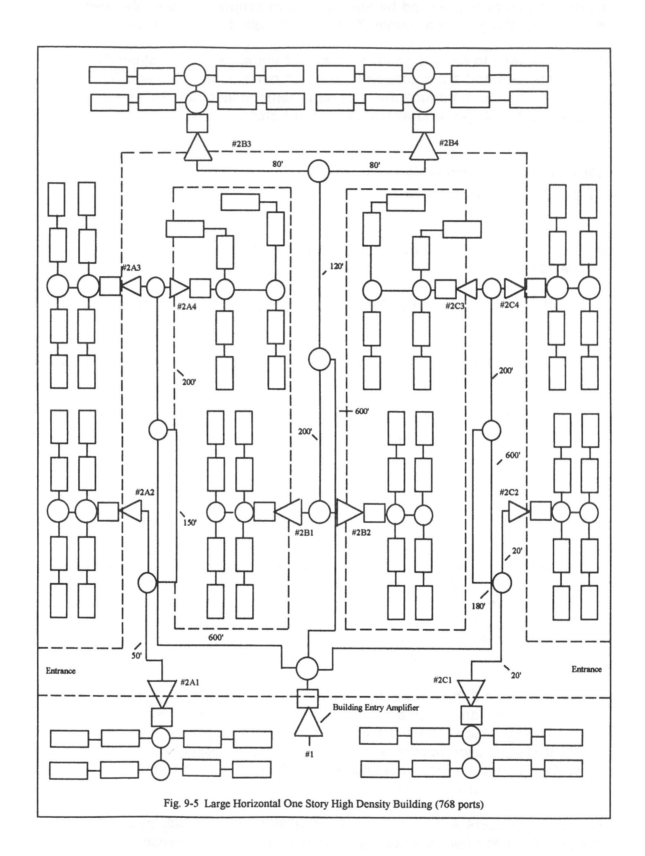

Fig. 9-5 Large Horizontal One Story High Density Building (768 ports)

218

splitter, 500 ft. of cable, a 2-way splitter, a 200 ft. cable and another 2-way splitter. Due to installation requirements, the last 200 ft. are changed to 150 ft. plus 50 ft., and in riser C to 180 ft. and 20 ft. As a result, all input levels to all 12 secondary amplifiers will be identical. For 750 MHz, the input level will be +10.88 dBmV using 500 JCAR cable. The output level of both 1st and 2nd amplifier will be +43.0 dBmV.

All amplifier areas are identical. A total of eight 8-port multitaps are connected to each amplifier. Assuming a maximum cable footage of 80 ft. between the splitter and the last tap, the tap-port level at 750 MHz is +43.0 dBmV, minus 4/29 tap insertion of 1.2 dB, minus 80 ft. of cable of 1.36 gives us a level into the tap of +33.24, which qualifies for a 17-value tap to meet our port-level criteria of +13.7 to +19.7 dBmV. The second tap should then be a 15-value tap. Using an uneven 3-way splitter with two high-loss and one low-loss port at the amplifier and by connecting the low-loss port to the splitter, all attenuations between amplifier and first multitaps are identical for every one of the 12x4 or 48 multitaps in the building. Different cable lengths to the first tap is the only factor that can alter slightly this perfect symmetry.

The Riser Design Process
for Transmissions in the Return Direction

Return Transmission in Building No. 158

We have previously determined that a modulator, when connected to the transmit outlet port, will be set to an output level of +55.0 ±2.0 dBmV.

All return transmission levels, after passing through the 150 ft. service drop, will enter the multitap port at the following averaged levels:

- at lowest frequency (5 MHz): +50.0 ±2.0 dBmV
- at highest sub-low frequency (46 MHz): +48.0 ±2.0 dBmV
- at highest high-split frequency (180 MHz): +46.0 ±2.0 dBmV

Looking at Fig. 9-2, the final Riser Design for Building No. 158, we find that we have optimized the placement of components for forward transmission levels.

Will the forward design of Fig. 9-2 provide acceptable return transmission characteristics?

To answer this question, we have to calculate the return transmission levels for each of the three branches.

It is our goal to arrive at the input of the building-entry amplifier with a level of +17.0 ±3.0 dBmV for each of the frequencies. If we can do this, then all return transmissions, whether through coaxial cables or through fiber-optic cables, will arrive at the headend in a ±3.0 dB window.

Building No. 158 - Branch #1 - Return Transmission

Since every tap can send a return signal, we have to conduct as many return calculations as we have multitaps. In branch #1 we have two multitaps, #5A and #5B. Our calculation then starts with tap #5B and seeks to determine what the return transmission levels are for 5, 46 and 180 MHz into the 3-way splitter port:

	5 MHz	46 MHz	180 MHz
Tap-port level	+50.0	+48.0	+46.0
Tap loss - 4/8 tap	-8.0	-8.0	-8.0
20 ft. of 500-JCAP	-0.034	-0.116	-0.248
Insertion 8/15 tap	-3.2	-3.0	-3.5
130 ft. of 500-JCAP	-0.221	-0.754	-1.612
Level at 3-way splitter	+38.54	+36.13	+32.64

The next location to calculate return levels from is the 8/15 tap, tap #5A:

	5 MHz	46 MHz	180 MHz
Tap-port level	+50.0	+48.0	+46.0
Tap loss - 8/15 tap	-15.0	-15.0	-15.0
130 ft. of 500-JCAP	-0.221	-0.754	-1.612
Level at 3-way splitter	+32.25	+32.25	+29.39

What we are finding is an about 2 dB range between 5 and 46 MHz signals and about a 5 to 6 dB range between 5 to 180 MHz signals. We have to calculate the other branches first before we can make a judgment.

Building No. 158 - Branch #2 - Return Transmission

In this branch we have four multitaps, so we need four calculations. To ease the task, we start with the closest tap:

	5 MHz	46 MHz	180 MHz
Tap-port level #4	+50.0	+48.0	+46.0
Tap loss - 4/20 tap	-20.0	-20.0	-20.0
20 ft. 500-JCAR	-0.032	-0.104	-0.2
Level at 3-way splitter	+29.97	+27.9	+25.8
Tap port level #3	+50.0	+48.0	+46.0
Tap loss 4/17 tap	-17.0	17.0	17.0
20 ft. 500-JCAR	-0.032	-0.104	-0.2
Insertion 4/20	-0.5	-0.7	-0.8

| 20 ft. 500-JCAR | 0.032 | -0.104 | -0.2 |
| Level at 3-way splitter | +32.44 | +30.09 | +27.8 |

	5 MHz	46 MHz	180 MHz
Tap-port level #2	+50.0	+48.0	+46.0
Tap loss - 8/15 tap	-15.0	-15.0	-15.0
20 ft. 500-JCAR	-0.032	-0.104	-0.2
Insertion 4/17	-0.9	-0.9	-1.1
20 ft. 500-JCAR	-0.032	-0.104	-0.2
Insertion 4/20	-0.5	-0.7	-0.8
20 ft. 500-JCAR	-0.032	-0.104	-0.2
Level at 3-way splitter	+33.05	+31.09	+28.5

	5 MHz	46 MHz	180 MHz
Tap-port level #1	+50.0	+48.0	+46.0
Tap loss - 8/11 tap	-11.0	-11.0	-11.0
20 ft. 500-JCAR	-0.032	-0.104	-0.2
Insertion 8/15	-3.2	-3.0	-3.5
20 ft. 500-JCAR	-0.032	-0.104	-0.2
Insertion 4/17	-0.9	-0.9	-1.1
20 ft. 500-JCAR	-0.032	-0.104	-0.2
Insertion 4/20	-0.5	-0.7	-0.8
20 ft. 500-JCAR	-0.032	-0.104	-0.2
Level at 3-way splitter	+34.27	+31.98	+28.8

What we readily see is that levels at 180 MHz are within 3 dB of each other. The same holds true for 46 MHz. Variations at 5 MHz, however, are more than 4 dB, which is caused by the inaccuracies of stated insertion losses at low frequencies.

Building No. 158 - Branch #3 - Return Transmission

To get the full picture, we should complete the branch #3 calculations:

	5 MHz	50 MHz	180 MHz
Tap-port level #5	+50.0	+48.0	+46.0
Tap loss - 4/17 tap	-17.0	-17.0	-17.0
Level at 3-way splitter	+33.0	+31.0	+29.0

	5 MHz	50 MHz	180 MHz
Tap-port level #6	+50.0	+48.0	+46.0
Tap loss - 4/17 tap	-17.0	-17.0	-17.0
20 ft. 500-JCAR	-0.032	-0.104	-0.2
Insertion loss 4/17	-0.9	-0.9	-1.1
Level at 3-way splitter	+32.07	+30.0	+27.7

At this point, we know the level of any return transmission arriving at the input to the 3-way splitter in the 5th floor. We still have to bring these signals down to the amplifier.

Building No. 158 - Return Transmission Levels
at the Amplifier Input

First, we have to determine the attenuation of the 3-way splitter and the 170 ft. riser cable:

	5 MHz	50 MHz	180 MHz
3-way splitter	-6.3	-5.7	-5.75
170 ft. 500-JCAR	-0.27	-0.88	-1.7
Attenuation add-on	-6.57	-6.58	-7.45

By adding these attenuations to all levels, formerly calculated for the input to the 3-way splitter, we can summarize the building No. 158 return transmission levels at the input to the reverse amplifier:

Level at Amp. Input	5 MHz	50 MHz	180 MHz
Branch 1 - Tap #5B	31.9	29.55	25.19
Branch 1 - Tap #5A	28.21	25.67	21.94
Branch 2 - Tap #4	23.4	21.32	18.35
Branch 2 - Tap #3	25.87	23.51	20.35
Branch 2 - Tap #2	26.48	24.51	21.05
Branch 2 - Tap #1	27.7	25.4	21.35
Branch 3 - Tap #5	26.43	24.42	21.55
Branch 3 - Tap #6	25.5	23.42	20.25

An averaging process would result in the following input levels:

26.94	24.73	21.25

Building No. 158 - Return Transmission Summary

These results tell us the following:

- When we design for a sub-low return system that works between 5 and 42 MHz, we can expect the return amplifier input levels to be within 3 dBmV. All we need to do is to plug in an 8 dB pad to compensate for the levels above +17.0 dBmV. Equalizers are barely required and should be used to preemphasize the amplifier's output level.

- When we design a high-split system that works between 5 and 180 MHz, both equalization and level control are important. The pads and equalizers are in the amplifier output section and have to control level and slope differences at the input to the next amplifier. In addition, they are used to preemphasize the amplifier output signal so that matching inputs are assured at the next amplifier.

- The variations within levels at each of the design frequencies of 5, 42 and l8O MHz are a little excessive. We are finding a level difference of over 8.5 dBmV between tap #5B and #4 at 5 MHz. At 42 MHz, we find an 8.23 dBmV difference between the same two multitaps and at l8O MHz, we have a 6.84 dBmV difference. The reasons of this variation are plenty -

- we had to use a 4/2O tap in branch #2 to make it all the way to the first floor

- the tap value selection of #5B is marginal. We could use a 4/ll tap to favor the return transmission characteristics a little more

- but most of all, the equipment layout is center-fed but not perfectly symmetrical. Each branch has different cable lengths and number of taps. In hindsight, we could improve the Fig. ll-2 building No. l58 design by bringing the riser to the 4th floor using a directional coupler for branch #2 taps (3 only). Use a splitter in the 5th floor from branch #l and have three taps in each direction

But, taking all these improvements into consideration, we still have not produced total symmetry. The l3O ft. of 5OO plenum cable will negate our efforts towards perfection.

Return Transmission in the High-Rise Building

Care has been taken in the forward design to arrive at a symmetrical equipment layout. The three amplifiers on floors 4, 9 and l4 serve exactly the same number and values of multitaps. The riser distribution from the building-entry amplifier center-feeds the amplifier at the 9th floor and uses identical cable lengths to feed the other two amplifiers. The entire distribution network is perfectly symmetrical.

When we calculate the return transmission levels, we only have to check one of the three amplifier areas since they are all identical.

In the riser, we only have to compare the 9th floor amplifier with one of the other two. Even the equipment layout for floors Bl and l is equal to the other floors, so no checking is required.

Fig. 9-4 provides us with both forward and return amplifier levels and indicates the perfect symmetry of the system. But let us do the calculations first. Again, as before, we need a calculation for each tap on each floor.

Step I - Return Transmission from the 8/I5 Tap on the I6th Floor

	5 MHz	5O MHz	I8O MHz
Tap-port level	+5O.O	+48.O	+46.O
Tap loss - 8/I5 tap	-I5.O	-I5.O	-I5.O
I5 ft. 5OO-JCAR	-O.O24	-O.O78	-O.I5
2-way Splitter	-3.8	-3.7	-3.75
I5 ft. 5OO-JCAR	-O.O24	-O.O78	-O.I5
3-way splitter	-6.3	-5.7	-5.75
Level into I4th floor Amp.	+24.85	+23.44	+2I.2

Step 2 - Return Transmission from the 4/8 Tap on the I6th Floor

	5 MHz	5O MHz	I8O MHz
Tap-port level	+5O.O	+48.O	+46.O
Tap loss - 4/8 tap	-8.O	-8.O	-8.O
Insertion - 8/I5	-3.2	-3.O	-3.5
I5 ft. 5OO-JCAR	-O.O24	-O.O78	-O.I5
2-way Splitter	-3.8	-3.7	-3.75
I5 ft. 5OO-JCAR	-O.O24	-O.O78	-O.I5
3-way Splitter	-6.3	-5.7	-5.75
Level into I4th floor Amp.	+28.65	+27.44	+24.7

The calculation just completed is also correct for floor I2.

Step 3 - Return Transmission from the I5th Floor

The only difference between floors I5 and I6 is a I5 ft. section of cable. Since we cannot worry about fractional dB levels, we do not need to calculate anything.

This means that the Step I calculation for the 8/I5 tap applies to the 8/I5 taps at floors I6, I5, I2 and I3. The Step 2 calculation for the 4/8 tap applies also to the 4/8 taps at floors I6, I5, I2 and I3.

Step 4 - Return Transmission from the I4th Floor

Since the tap values are different on this floor, we should take a quick check first for the 8/2O tap:

	5 MHz	5O MHz	I8O MHz
Tap-port level	+5O.O	+48.O	+46.O
Tap loss - 8/2O	-2O.O	-2O.O	-2O.O
3-way Splitter	-6.3	-5.7	-5.75
Level into I4th floor Ampl.	+23.7	+22.3	+2O.25

and then for the 4/7 tap:

	5 MHz	50 MHz	180 MHz
Tap-port level	+50.0	+48.0	+46.0
Tap loss - 4/I7	-I7.0	-I7.0	-I7.0
Insertion - 8/20	-I.0	-I.0	-I.2
3-way Splitter	-6.3	-5.7	-5.75
Level into I4th floor Ampl.	+25.7	+24.3	+22.05

This level is identical on all three amplifier floors and the levels obtained in Steps I, 2 and 3 also apply to other floors. This level also applies to the basement and floor I, since we only have a 3-way splitter in the path.

Can we now determine the pad and equalizer values for the amplifiers?

Before we select pads and equalizers for the secondary amplifiers on floors 4, 9 and I4, it is necessary to determine the attenuation in the riser network to the building-entry amplifier.

Return Transmission in the Vertical Riser and Determination of Pads and Equalizers

The only difference between the three amplifiers is that the amplifier at floor 9 is not connected through a 75 ft. cable section. This is a minor difference of less than I dB. Again, one calculation will suffice for all three signal paths.

Table 9-5 gives us the amplifier output levels for 5 to 50 and 5 to I86 MHz amplifiers. So, depending upon the system that you are designing, the calculation has a few minor variances.

	Sub-Split		High-Split	
	5 MHz	50 MHz	5 MHz	180 MHz
Amp. output level	+.32.0	+32.0	+37.0	+37.0
75 ft. 500-JCAR	-0.I2	-0.39	-0.I2	-0.75
3-way Splitter	-6.3	-5.7	-6.3	-5.75
200 ft. 500-JCAR	-0.32	-I.I2	-0.32	-2.0
3-way Splitter	-6.3	-5.7	-6.3	-5.75
Test tap 4/29	-0.3	-0.5	-0.3	-0.5
Bldg. entry Amp.	+I8.66	+I8.59	+23.66	+22.25

The 5 to 50 MHz sub-split amplifier has a gain of I5 dB. The 5 to I80 MHz high-split amplifier has a gain of 20 dB. Whatever the signal level is at the input to the amplifier, it will be I5 or 20 dB higher at the output.

We now have to analyze the previously calculated amplifier inputs to determine what pads and equalizers we need to bring all signals to the building-entry amplifier with a level of +17.0 ±3.0 dBmV.

It is best to tabulate the calculated information so that the level variations are easily detectable. Looking at the 14th floor amplifier inputs should suffice.

Amp. Inputs	5 MHz	50 MHz	180 MHz
8/15 from floor 16	+24.85	+23.44	+21.2
4/8 from floor 16	+28.65	+27.44	+24.7
8/15 from floor 15	+24.85	+23.44	+21.2
4/8 from floor 15	+28.65	+27.44	+24.7
8/20 from floor 14	+23.7	+22.3	+20.25
4/17 from floor 14	+25.7	+24.3	+22.05
8/15 from floor 13	+24.85	+23.44	+21.2
4/8 from floor 13	+28.65	+27.44	+24.7
8/15 from floor 12	+24.85	+23.44	+21.2
4/8 from floor 12	+28.65	+27.44	+24.7

The average level   at 5 MHz   is +26.3
                          at 50 MHz is +25.0
and   at 180 MHz is +22.6

Using these averages, we can now determine the pad values of the three riser amplifiers.

In a sub-split system, we need an 8 or 9 dB pad to bring the signal closer to a desirable input level of +17.0 dBmV. But, we also have to consider the riser level into the building-entry amplifier. That level was around +18.6 dBmV. So, we need about a 10 dB pad to obtain an about +17.0 dBmV input. Let us follow the signal level for both sub-split and high-split configurations:

	5 MHz	50 MHz	5 MHz	180 MHz
Aver. inp. to 14th fl. amp.	+26.3	+25.0	+26.3	+22.6
Amp. gain	+15.0	+15.0	+20.0	+20.0
Total output level	+41.3	+40.0	+46.3	+42.6
Loss in riser	-13.34	-13.41	-13.34	-14.75
Level without pad	+27.96	+26.59	+32.96	+27.85
Required pad value	10 dB		11 dB	
Inp. level to 1st amp.	+17.96	+16.59	+21.96	+16.85

The pad values for amplifiers at the 14th, 9th and 4th floor are plug-in types and attenuate the output of the amplifier. In the sub-split system, the equalizer value can be 0 dB. In the high-split system, we need a 4.5 or 6 dB equalizer to produce a flat signal input to the building-entry amplifier.

Return Transmission in the Large Horizontal Building

A review of the hypothetical riser design of the large horizontal building in Fig. II-5 shows a very symmetrical center-fed layout of all network equipment. As in the high-rise building, symmetrical design increases the ability to keep all return transmission levels within close tolerances.

To a great extend, the amount of care given towards symmetry while designing the forward transmission system is a good indicator for the quality of the return transmission.

As it was recommended before in the forward transmission design section, and as it was proven in the forward design of the high-rise building, it is important to

- cluster groups of multitaps with up to 60 ports
- center-feed the cluster with an amplifier
- interlink all amplifier areas with a symmetrical riser feed that originates from the MDF building-entry amplifier

It is, of course, mandatory to calculate at least one amplifier area and one riser distribution network to determine amplifier input levels and to assure that all secondary amplifiers are padded and equalized properly. Only then can a +17.0 ±2.0 dBmV flat level be assured at the input of the building-entry amplifier.

Fig. 9-5 shows 12 amplifier areas with eight 8-port taps each, for a total of 768 ports. All 12 amplifiers are secondary amplifiers and in parallel. There are only two amplifiers in cascade. The riser feed points have been carefully chosen to center-feed a group of 4 amplifiers. The tap values of all first and all second multitaps are the same. It may be a good exercise to calculate both forward and return of one amplifier area and of the riser network. Assuming short distances (20 ft.) between taps, you will obtain perfectly matched outlet levels and you will find return levels at the building-entry amplifier to be within 17.0±1.5 dBmV.

The Design Documentation

A comprehensive design documentation helps to control the system implementation relative to the Bill-of-Materials and the test data.
Taking the high-rise building of Fig. 9-4 as an example, the design documentation should consist of

- forward port and outlet levels
- forward amplifier data

- return amplifier data
- Bill-of-Materials

Table 9-8 shows a tabular presentation of material requirements. The table contains equipment types and quantities, cable types and quantities as well as connector types and quantities. Every device needs a housing connector to terminate the cable, or a housing-to-housing connector when two devices are adjacent to each other.

Fl.	Amp. No.	Riser (ft)	2way Split	3way Split	Hsng Con	Hsng to Hsng	8/15	8/20	4/8	4/17	2×2 Wall Brd.	4×4 Wall Brd.
16		15			1	1	1		1		1	
15		15	1		2	2	1		1		1	
14	2C	30		1	2	3		1		1		1
13		30	1		2	2	1		1		1	
12		15			1	1	1		1		1	
11		30			1	1	1		1		1	
10		30	1		2	2	1		1		1	
9	2A	45		1	5	4		1		1		1
8		45	1		2	2	1		1		1	
7		30			1	1	1		1		1	
6		45			1	1	1		1		1	
5		45	1		2	2	1		1		1	
4	2B	30		1	2	3		1		1		1
3		30	1		2	2	1		1		1	
2		15			1	1	1		1		1	
1		30			1	1		1		1	1	
B1	1	65			2	2		1		1		1
B2												
Total		545	6	3	30	31	12	5	12	5	13	4

Table 9-8 - Bill-of-Materials - High-rise Building

The cable footage of Table 9-8 should be increased for slippage, dressing and connectorization. A 20% increase is recommended.

Table 9-9 shows the calculated tap-port levels for every floor of the high-rise building.

228

Floor	8 port 50MHz (dBmv)	8 port 220MHz (dBmv)	8 port 750MHz (dBmv)	4 port 50 MHz (dBmv)	4 port 220 MHz (dBmv)	4 port 750 MHz (dBmv)
16	11.96	15.56	14.06	15.96	16.06	16.96
15	11.96	15.56	14.06	15.96	16.06	16.96
14	10.8	11.7	14.6	12.8	13.5	15.8
13	11.96	15.56	14.06	15.96	16.06	16.96
12	11.96	15.56	14.06	15.96	16.06	16.96
11	11.96	15.56	14.06	15.96	16.06	16.96
10	11.96	15.56	14.06	15.96	16.06	16.96
9	10.8	11.7	14.6	12.8	13.5	15.8
8	11.96	15.56	14.06	15.96	16.06	16.96
7	11.96	15.56	14.06	15.96	16.06	16.96
6	11.96	15.56	14.06	15.96	16.06	16.96
5	11.96	15.56	14.06	15.96	16.06	16.96
4	10.8	11.7	14.6	12.8	13.5	15.8
3	11.96	15.56	14.06	15.96	16.06	16.96
2	11.96	15.56	14.06	15.96	16.06	16.96
1	10.8	11.7	14.6	12.8	13.5	15.8
B1	10.8	11.7	14.6	12.8	13.5	15.8
B2						

Table 9-9 - Calculated Tap-port Levels - High-rise Building

Table 9-IO compiles the amplifier data in both forward and reverse direction.

Forward Amplifier Levels

Floor	Amp. No.	Input 50MHZ dBmv	Input 220Mhz dBmv	Input 750MHz dBmv	Pad (dB)	Equalize (dB)	Output 50MHz dBmv	Output 220MHz dBmv	Output 750MHz dBmv
14	2C	+23.67	+22.85	+21.46	9.0	6.0	+37.0	+38.0	+43.0
9	2A	+24.06	+23.86	+23.08	10.0	9.0	+37.0	+38.0	+43.0
4	2B	+23.67	+22.85	+21.46	9.0	6.0	+37.0	+38.0	+43.0
B1	1*	—	—	—	—	—	+37.0	+38.0	+43.0

* part of outsid- plant design

Return Amplifier Levels

Floor	Amp. No.	Input Av. 5MHz dBmv	Input Av. 50MHz dBmv	Input Av. 180MHz dBmv	Pad sub/ high dB	Equal. sub/ high dB	Output 5MHz sub/high dBmv	Output 50MHz dBmv	Output 180MHz dBmv
14	2C	26.3	25.0	22.6	10/11	0/6.0	31.3/36.3	30.0	31.6
9	2A	26.3	25.0	22.6	10/11	0/6.0	31.3/36.3	30.0	31.6
4	2B	26.3	25.0	22.6	10/11	0/6.0	31.3/36.3	30.0	31.6
B1	1	17.96/ 21.96	16.59	16.85	*/*	*/*	32.0/37.0	37.0	37.0

Table 9-IO - Amplifier Forward and Return Levels

Having a valid Bill-of-Material and a comprehensive documentation of forward and return calculations is the first step towards system implementation.

In the event you are planning the issuance of an RFP, remember that since you have good information, you may also consider to issue even an IFB format. In any event, do not fail to make the bidder responsible for final design verification and provision of whatever material is required to complete the system.

Whatever your plans - you have collected the information in great detail - you have optimized the design by careful analysis of center-feeding and riser symmetry and you are now in the position to supervise every work activity of an implementor and obtain the highest quality of performance from your broadband system design concept.

Chapter 10

The HFC Broadband Network Design Process - Outside-Plant Design -

The Hypothetical Campus Layout

Fig. 7-3 showed a hypothetical campus layout that we will use again to illustrate the outside-plant design process.

In the Outside Plant information section of Chapter 7 - Design Information Checklist - we already collected outside-plant components such as Aerial Pole Line data and Existing Conduit data.

In addition, we analyzed the conditions that affect the decision to use fiber-optic cable vs. coaxial cable.

Optimizing the HFC System

Optimizing the Hypothetical Campus HFC System

The optimization process of an HFC transmission system is primarily a function of the number of users and the distance experienced in a group of buildings. Using common sense as the main criteria, the optimized HFC system for our hypothetical campus of Fig. 7-3 may look as follows:

a)   Coaxial service for building No. 132 and buildings No. 128, 140, 120 and 148. These buildings are adjacent to the video operations center and have only a total of 102 outlets. No more than one amplifier is used per building. All amplifiers can be in parallel

b)   A fiber-node for building R100 with 192 outlets, requiring 2 amplifiers in cascade within the building

c)   A fiber-node for building R110 with 192 outlets, requiring 2 amplifiers in cascade within the building

d)   A fiber-node for building R150 with 180 outlets, requiring 2 amplifiers in cascade within the building

e)   A fiber-node for building No. 158 with coaxial distribution for buildings No. 152 and 180. There will be only 1 amplifier required per building and all amplifiers can be in parallel. There are only 87 outlets. When the new conduit is constructed, building No. 182 should be included in this node

f)   A fiber-node for building R170 with coaxial distribution for buildings R160, R170, 184 and possibly 182. This node would consist of a maximum of 208 outlets. The coaxial plant in R170 would consist of 2 amplifiers in cascade. No more than 2 amplifiers in cascade would be required for buildings R160, 184 and 182

This optimized HFC system would use 5 fiber-nodes. The crossection of the cable requirement between building No. 132 and MH-A has been reduced to five 12-strand fiber-optic cables and one coaxial cable. The longest fiber cable distance is to building R170 for 1,958 ft.

Fiber-optic cables are used between

-   Building No. 132 and R100
-   Building No. 132 and R110

- Building No. 132 and R150
- Building No. 132 and R170
- Building No. 132 and 158

Coaxial cables are used between

- Building No. 132 to 128, to 140, to 120
- Building No. 132 and 148
- Building No. 158 to MH-A, to 152 and 180
- Building R170 to MH-E and to R160, 182 and 184

Summary of HFC Trade-off Considerations

As we already discussed in previous chapters, the decision to use fiber-optic cables over coaxial cables must make sense from an economical, practical, political and forward-looking point of view.

Distance Considerations

Depending upon the size of your campus, the use of coaxial outside plant should be limited to about 2500 ft. If you find that the distances between the headend or video operations center are in excess of 2500 ft., a single-mode fiber cable to a convenient fiber-node location is recommended. This maximum distance is determined by the desire to keep the number of cascaded amplifiers under four and assumes that two amplifiers in cascade will be required within the building. Outside-plant distances of less than 2500 ft. assure that no more than two amplifiers in cascade are used.

Service Density Considerations

If your campus contains buildings with a potential for more than 150 outlets, then a fiber-node may be a practical consideration. In general, a fiber-node should be located to serve not more than 500 outlets and not less than 150.

Conduit Availability Considerations

Review your available conduit space between the headend and every building, and group of buildings. Keep in mind that a coaxial system leaving the headend consists of one cable for both forward and return transmissions. The diameter of this coaxial cable will be between 0.5 and 0.86 inches.

When using fiber only, a one-way transmission is achieved. One fiber is required for forward transmission and one or even multiple fibers are required for the return transmission. So, it is easy to use a lot of conduit space when routing fiber to the headend location.

When conduits are filled to capacity and construction is required to accommodate multiple fiber-optic cables, the question should be asked "can I avoid new conduit construction by using a single coaxial cable?"

Return Transmission Considerations

Return transmission from any outlet of the network is accomplished by standard RF modulators. The baseband video/audio or a digital bit-stream are modulated to a carrier frequency in the return spectrum of the coaxial cable. When the coaxial cable extends all the way to the headend, no further signal translation is required.

When the coaxial cable, however, ends at a fiber-node, all channels have to be either

- demodulated for baseband transmission on multiple fiber

- or it has to be forwarded to a multi-channel fiber transmitter that can handle the entire 5 to 186 MHz return band

Neither solution is very cost-effective. Using a single fiber for every return signal can require as many fiber-optic cables as the number of return outlets to produce a nonblocking return transmission path that is available to every location at any time. In addition, demodulation equipment is required.

Transmitting the entire return transmission band requires multi-channel fiber-optic transmitters that become more expensive with the number of channels that are transmitted.

For this reason, it is recommended to

a)   use coaxial cable for the return between fiber-nodes and the headend.
     At an attenuation of 0.6 dB per 100 ft. at 180 MHz, a distance of 4500 ft. can be covered using an 860 CATV coaxial cable without requiring any amplifier

The major benefit of this approach is the nonblocking nature of this architecture. Return transmissions can be made from any outlet and at any time.

b)   determine the maximum number of simultaneous return transmissions through the fiber-node location. Whatever the number is, it must be individual fiber strands between fiber-node headend or can be used

234

to stack return channels in groups of 16 or 24 for transmission on the fiber

The major problem of this approach is the limitations of the number of return transmissions that can be accommodated at the same time as well as the substantially higher costs.

The Broadband Coaxial Outside-Plant Segments

The Forward Transmission Design

A review of the hypothetical campus of Fig. 7-3 shows that there are three separate coaxial cable areas.

Area I combines the buildings closest to the assumed headend location in Building 132. Area 2 interconnects Buildings 152 and 180 with the fiber-node location in Building 158 and Area 3 connects Buildings R160, 182 and 184 with the fiber-node in Building R170.

Before we can calculate levels, we have to determine the type of cable and amplifiers that we want to use.

Cable Selection

It can be concluded from the listing of cable types in Chapter 8 that the choice of outside-plant cable types is plentiful. Manufacturing coaxial cables for millions of miles of Cable TV distribution systems has matured this industry.
Cables come in various sizes and very much reflect the trend of the industry to overcome distances by using a larger diameter cable.

The physical properties of all cables have gone through an evolutionary process during the past two decades and are able to withstand wind, weather, vibrations and longitudinal stress for 20 years or more.

The shielding properties are 100% since all feature a solid aluminum sheath, a requirement to keep ingress and egress confined to manmade connectorization problems.

The maximum pulling tension of all coaxial cables with a diameter over 0.5 inch is in excess of 400 lbs., which compares well to fiber-optic cabling with 12 strands or more.

The only real difference between the cables is the attenuation per 100 ft., which, of course, is a direct function of the size of the cable. The more air can be found between the center conductor and the shield, the lower the attenuation.

Table 10-1 shows a comparison of cable types, diameter and attenuation at the prevailing forward and return transmission frequencies.

Cable Type	Diamtr. (in)	5 MHz dB/100ft	50 MHz dB/100ft	180 MHz dB/100ft	220 MHz dB/100ft	750 MHz dB/100ft	Pull Tension (lbs)
500JCA	0.56	0.16	0.52	1.0	1.11	2.16	300
625JCA	0.685	0.13	0.42	0.85	0.94	1.78	475
750JCA	0.82	0.11	0.35	0.68	0.76	1.48	675
875JCA	0.945	0.09	0.30	0.6	0.67	1.29	875
QR540 JCA	0.61	0.14	0.44	0.88	0.98	1.85	220
QR860 JCA	0.96	0.09	0.30	0.59	0.65	1.24	450

Table 10-1 - Attenuation of Outside-Plant Cables

Looking at a typical spacing between trunk amplifiers of 22 dB, we can determine what 22 dB means with respect to the distance. A 500-JCA cable lets us put the amplifiers apart by 22÷2.16/100 ft. equals 10.18 ft.times 100 ft. or 1,018 ft. A QR 860 JCA cable permits us to span more distance. 22÷1.24/100 ft. equals 1,774 ft.

Usually, a campus environment is not big enough to justify low-attenuation cables. Buildings are in close proximity. It is, therefore, more important to focus on a good pulling tension and find a good compromise relative to attenuation.

From a review of the above table, it seems very appropriate to select the 750 JCA cable. With a maximum pulling tension of 675 lbs., we can even pull it through the already congested duct system. And with an attenuation of 1.48 dB per 100 ft., we can span 1,500 ft. at 750 MHz. At 180 MHz, we can even go over 3,200 ft.This may come in handy to establish an economical return transmission system.

Amplifier Selection

Any coaxial-cable outside plant should be looked upon as a trunkline. Coaxial broadband technology distinguishes between trunk and distribution areas. The distribution portion is the area that provides services to the subscribers. The trunk area interconnects the various distribution areas.

Distribution amplifiers like the ones that we have used for in-building distribution, have high-output levels to feed multitaps. A high output-level amplifier produces a higher level of intermodulation product such as crossmodulation and triple-beat products. For this reason, no more than 2 amplifiers can be used in the distribution area and a derating of 3 dB has become standard practice.

Trunk amplifiers, on the other hand, have a relative low gain of about 22 dB, or 26 dB gain, and operate at a low output level of about 33 dBmV. This method reduces intermodulation products and permits a cascade of many amplifiers.

The theory of the HFC system design does not permit more than two trunk amplifiers in cascade, a limitation that will reduce the number of electronic devices connected to a fiber-node. The lower the number of electronics, the lower the number of outages, the lower the cost of maintenance and the higher the quality of the video signals and the reliability of the system.

Our campus design standards call for not more than three amplifiers in cascade. Two already have been used and are reserved for distribution within the buildings. This leaves us with one trunk station to interconnect buildings following the fiber-node location. As we can see from a review of Fig. 10-1 and 10-2, this goal has been met for our hypothetical campus.

The choice of trunk amplifiers is quite limited. The electronics have matured throughout the evolutionary process of the cable industry.

The trunk amplifier requirements for forward transmission are listed below. The return transmission requirements are identical to the data listed in Table 9-5 entitled "Distribution Amplifier Requirements".

	50-750 MHz	222-750 MHz
Frequency Response	$\pm$0.25 dB	$\pm$0.25 dB
Minimum Full Gain	22 or 26 dB	22 or 26 dB
Noise Figure	9.5	9.5
Output Level	+33.0 dBmV	+33.0 dBmV
Tilt	3	3
Cross Modulation	91 dB	91 dB
Triple Beats	92 dB	92 dB
Typical Input Level	+11.0 dBmV	+11.0 dBmV

A comparison with the distribution amplifiers of Table 9-5 shows that the flatness is in a much higher range and intermodulation products are 30 dB lower.

Forward Level Calculations (Area I) - Service to Buildings 128, 140. 120 and 148

Before starting the calculations, it is good to overview the geography of the area and add up a cumulative distance to the last amplifier.

In our Service Area I, we start with the location of the headend. The headend is located in the 3rd floor of building 132. Approximately 50 ft. of cable are required to the branch-off point for building 128. Another 50 ft. of cable are required to link with the conduit-entrance point. The distances to building 148 are 300 ft. to MH-A and another 120 ft. plus 50 ft. inside. The total amounts to 570 ft. Using the 750 JCA cable with 1.48 dB/100 ft. at 750 MHz, we can determine that the total cable loss is 8.44 dB.

The other branch of the Area I system goes through buildings 128, 140 and into building 120. The MDF, or building entry location in building 120 is 50 ft. inside the building. The total cable distance of this branch then is 770 ft. This difference expressed in dBs at 750 MHz is 11.4 dB.

We also know that we have to split once in building 132 and we need 2 splitters to feed amplifiers in buildings 128 and 140 before we get the signal to building 120. This adds 4.9 and 7.2, or 12.1 dB, to the 11.4 dB of cable loss, for a total of 23.5 dB.

In building 132, we have to place a trunk amplifier. It has an output of +33 dBmV at 750 MHz. Deducting the 23.5 dB from the 33 dBmV tells us that the input level to the last amplifier (building 120) will be 9.5 dBmV. This value is a little too low because the distribution amplifier needs +13 dBmV at the input to produce a +46.0 dBmV output level.

What can be done, without adding another amplifier or having to change the cable size?

The splitter in building 132 can be changed to a directional coupler. A 12 dB directional coupler has a loss of 12 dB in one direction and 2 dB in the other. Using the low-loss end for the branch, going to building 128, we can improve the input level by 2.9 dB. The resulting input level for the amplifier is now +12.4 dBmV, which is low but acceptable.

But, by doing this, we have increased the loss in the building 148 branch by 12 dB. If we add this to the 8.44 dB of cable loss, we have a total loss of 20.44 dB, which, if deducted from +33.0 dBmV, gives us an input level of +12.56 dBmV, which just meets the minimum input requirement within 0.5 dB.

There is one more thing we can improve on. Remember the ground rule of symmetrical center-feeding. Well, we have three buildings to feed in the building

I28 branch. Why don't we center-feed the building I4O amplifier and, by using a 3-way splitter, have equal-distance cable lengths going to both the building I2O and I28 amplifiers.

So, before calculating each and everyone level at all devices and for every frequency, take a good look at the geography and optimize the system by making a few rough calculations at the upper design frequency of 75O MHz.

Fig. IO-I shows the final layout of components of Service Area I. It appears that in each of the buildings the building-entry amplifier will take care of the low number of outlets. This means that the maximum number of amplifiers in cascade is two.

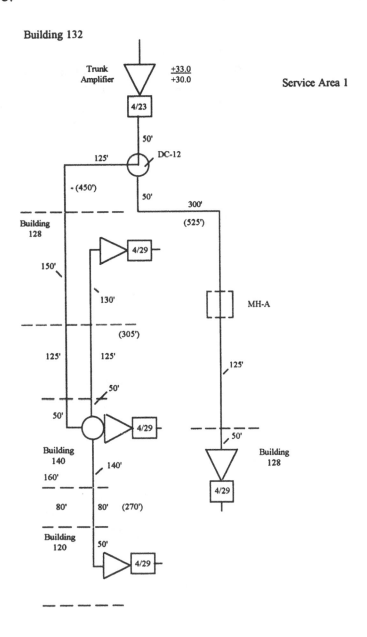

Fig. 10-1 Area 1 Coaxial Design

Step 1 - Service to Building 128, 140 and 120 Branch

We are now ready to make some detailed calculations:

	50 MHz	220 MHz	750 MHz
Trunk Amp. output	+30.0	+31.0	+33.0
50 ft. 750-JCA	-0.175	-0.38	-0.74
DC-12	-0.9	-1.1	-2.0
450 ft. 750-JCA	-1.575	-3.42	-6.60
3-way Splitter	-5.7	-5.8	-7.2
270 ft. 750-JCA	-0.95	-2.05	-4.0
Inp. level to Amp. (120)	+20.7	+18.25	+12.46
Inp. level to Amp. (140)	+21.65	+20.2	+16.46

It appears that, except for the amplifier in building 140, we are about 0.5 dBmV short of the desired input level. In the case of the building 128 amplifier, we have an additional 35 ft. of cable, which will take away another 0.52 dBmV from the desired +13.0 dBmV. A 1 dB too low input level is not a disaster. All levels should be looked upon to be valid within a ±1.0 dB window. The amplifier reserve gain will compensate for this 1 dB design deficiency.

Step 2 - Service to Building 148

Again, we have to start at the trunk amplifier in the headend.

	50 MHz	220 MHz	750 MHz
Trunk Amp. output	+30.0	+31.0	+33.0
50 ft. 750-JCA	-0.18	-0.34	-0.74
DC-12	-12.0	-12.0	-12.0
525 ft. 750-JCA	-1.84	-3.99	-7.77
Amp. input (148)	+15.98	+14.67	+12.49

The slope of this configuration is only 3.5 dB for a sub-split system. In the other branch, the slope is 8.24 dB. The longer cable length of the building 128 branch accounts for the difference. For building 140, the value is halfway in between.

Step 3 - Determination of Pads and Equalizers

Both pads and equalizers are determined by the input level and by the difference to the lowest passband frequency. Pad values are not different for sub-split or high-split systems, equalizers, however, are.

The pad is inserted into the input circuitry of the amplifier. It is its function to keep the output level within the specified limits. To determine a pad value, take the specification output level of +46 dBmV (single amplifier) or +43 dBmV (2 amplifiers in cascade) and subtract the calculated input level. If the difference is less than 33 dB, the pad is to make up the difference to 33 dB.

The pad values for the four amplifiers are listed below:

		minus			
	Outp. Level	Inp. Level	=33 dB Gain	+	Pad
Amp. at I28	+46.0	+I2.0	=34.0	+	0
Amp. at I40	+46.0	+I6.46	=30.54	+	2.0
Amp. at I20	+46.0	+I2.46	=33.54	+	0
Amp. at I48	+46.0	+I2.49	=33.5I	+	0

Equalizers are also used at the input of an amplifier. New state-of-the-art 750 MHz amplifier have a built-in tilt of up to 9 dB. All our previous calculations assumed a 6 dB tilt at the output of the building-entry amplifier. This means that, if the design tilt between high and low is less than 3 dB, only a 0 dB equalizer is required.

Here are the recommended equalizer values for the four amplifiers:

	50-750 MHz sub-split	220-750 MHz high-split
Amp. at I28	4.5-6.0 dB	3-4.5 dB
Amp. at I40	3.0-4.5 dB	0-I.5 dB
Amp. at I20	4.5-6.0 dB	3-4.5 dB
Amp. at I48	0-3.0 dB	0-3.0 dB

A good selection of equalizers is recommended when aligning the network. Equalization is subject to amplifier nonlinearities and reflection differences within connectors and terminations. The most practical way is to select the correct value during system alignment.

Forward Level Calculation (Area 2)
Service to Buildings I58, I52 and I80

The design calculations follow the system architecture shown in Fig. I0-2, Area 2 - Design.

Step I - Service to Building I58

The fiber-node location is the begin point of the coaxial Area 2 system. In many ways, the fiber-node constitutes a mini-headend and, therefore, requires a trunk amplifier to route the signal to the other buildings in the service area.

The first calculation then is the determination of the building I58 distribution amplifier input level.

	50 MHz	220 MHz	750 MHz
Trunk Amp. output	+30.0	+31.0	+33.0
4/29 Tap	-0.5	-0.5	-1.2
DC-16	-16.0	-16.0	-16.0
Input to Bldg. Amp.(158)	+13.5	+14.5	+15.8

This level is just the right level for the distribution amplifier at the building-entry location. A 3 dB pad and a 3 dB equalizer should be the first selection.

Step 2 - Service to Building 152

Again, starting with the trunk amplifier, the calculation takes all cable and devices into consideration for the distance between building 158 and 152.

	50 MHz	220 MHz	750 MHz
Trunk Amp. output (158)	+30.0	+31.0	+33.0
4/29 Tap	-0.5	-0.5	-1.2
DC-16 Insertion	-0.7	-0.9	-1.7
190 ft. of 750-JCA	-0.665	-1.444	-2.81
2-way Splitter	-3.7	-3.8	-4.9
260 ft. of 750-JCA	-0.91	-1.98	-3.85
Input to Bldg. Amp.(152)	+23.53	+22.38	+18.55

The input levels are just right. A 5 or 6 dB pad and a 4.5 or 6 dB equalizer will provide for the necessary adjustments.

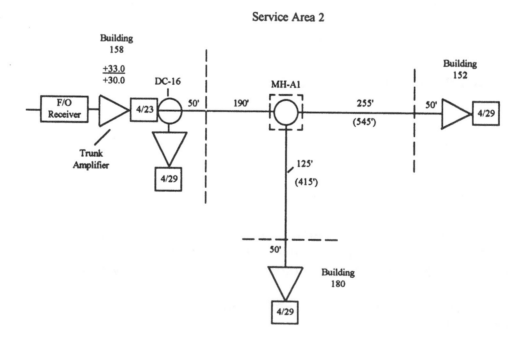

Fig. 10-2 Area 2 Coaxial Design

Step 3 - Service to Building 180

Was the choice of a 3-way splitter in manhole MH-Al correct, since the distance to building 180 is 130 ft. shorter?

A 130 ft. 750-JCA cable section has only an attenuation of 1.9 dB. So, we can conclude that, without making any changes, the input level to the building 180 distribution amplifier will be 1.9 dB higher than the building 152 level.

	50 MHz	220 MHz	750 MHz
Input level (152)	+23.53	+22.38	+18.55
Diff. 130 ft. less cable	+0.45	+0.99	+1.9
Input to Bldg. Amp.(180)	+23.98	+23.37	+20.45

These levels are manageable. The amplifier will need a 7 or 8 dB pad and a 3 or 4.5 dB equalizer to perform as desired. The Area 2 forward design is complete.

Forward Level Calculation (Area 3)
Service to Buildings R170, 160, 182 and 184

The design calculations follow the system architecture of Fig. 10-3, Area 3 - Design.

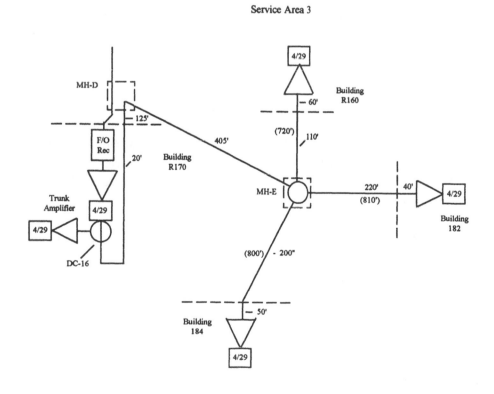

Fig. 10-3 Area 3 Coaxial Design

243

Step 1 - Service to Building RI7O

The geography of Area 3 is quite similar to Area 2. The trunk amplifier is followed immediately by the building distribution amplifier. Therefore, the same calculation that was made before for Area 2, building 158, applies.
The resulting input levels are again

	50 MHz	220 MHz	750 MHz
Input to Bldg. Amp.(RI7O)	+13.5	+14.5	+15.8

Again, a 3 to 4 dB pad and a 1.5 to 3 dB equalizer will adjust the levels to the desired values.

Step 2 - Service to Building 182

Fig. 10-3 shows as total cable footage between buildings RI7O and 182. This footage amounts to about 12 dB at 750 MHz. When adding the loss of the 3-way splitter in MH-E and the DC-16 at RI7O, total loss at 750 MHz is about 19 to 21 dB. Subtracting this from the trunk amplifier output (+33 dBmV) will provide an input level between 12 and 14 dBmV. It looks like this level may be marginal.

Is there an alternative Solution?

Changing the 3-way splitter to a combination of directional couplers does not change the relationship of the signals in MH-E. The use of an uneven 3-way splitter does not improve the levels either since the distance to both building 182 and 184 is about the same.

If the calculation would show an input level below +11 dBmV, or 2 dB lower than the recommended input level, the only alternative would be to place a second trunk amplifier in buildings 182 and 184. The number of service drops are low and require only one building amplifier. This means that there are two trunk stations and one distribution amplifier in cascade, which meets the three-amplifiers-in-cascade goal.

Another alternative is using a lower loss cable. The detailed design follows:

	50 MHz	220 MHz	750 MHz
Trunk Amp. outp.(RI7O)	+30.0	+31.0	+33.0
4/29 Tap	-0.5	-0.5	-1.2
DC-16 Insertion	-0.7	-0.9	-1.7
550 ft. of 750-JCA	-1.93	-4.18	-8.14
3-way Splitter	-5.7	-5.8	-7.2
260 ft. of 750-JCA	-0.91	-1.98	-3.85
Input to Bldg. Amp.(182)	+20.26	+17.64	+10.91

244

The resulting input level is over 2 dB out of spec. and the use of a larger cable size appears to be the most cost-effective solution.

Using 875 JCA, the total 810 ft. of cable attenuation is 10.45 dB at 750 MHz as opposed to the 11.9 dB in the above table (8.14 + 3.85). This approach improves the input level by about 1.5 dB, just enough to operate the system at safe levels.

A recalculation of the above table, using the larger size 875 JCA cable, results in the following input levels:

	50 MHz	220 MHz	750 MHz
Diff. due to cable size	+0.41	+0.74	+1.5
Input to Bldg. (182)	+20.67	+18.38	+12.41

A 0 dB pad and a 6 or 7.5 dB equalizer are required.

Step 3 - Service to Building 184

As we can see in Fig. 10-3, the footage difference between building 182 and 184 is 10 ft. No calculation is necessary. Using the same 875 JCA in all of Area 3, the input levels for the building 184 amplifier will be

	50 MHz	220 MHz	750 MHz
Input to Bldg. (184)	+20.65	+18.35	+12.36

A 0 dB pad and a 6 or 7.5 dB equalizer are required.

Step 4 - Service to Building R160

This branch is 90 ft. shorter than the distance to building 182. All that needs to be done is to determine the loss differential and record the design input levels.

	50 MHz	220 MHz	750 MHz
Diff. due to length	+0.27	+0.6	+1.16
Input at Bldg. (182)	+20.67	+18.38	+12.41
Input at Bldg. (R160)	+20.94	+18.98	+13.57

In this case, we may even need a 1 dB pad. An equalization of 6 to 7.5 dB appears necessary and will be equal to the values for the other building-entry amplifiers.

The Return Transmission Design

Return Level Calculations (Area I)
Service from Building 128, 140, 120 and 148

Return transmission from any location on a coaxial tree system requires that all levels arrive at the trunk amplifier in a window of +17.0 ±1.5 dB.

The somewhat symmetrical architecture of Area I should be able to accomplish this by varying pad and equalizer values in the output of each return amplifier.

To accomplish this, we can calculate against the signal flow and in the forward direction. For instance, we know that the trunk amplifier input level to the reverse amplifier wants to be +17.0 dB. By adding the losses of cable and passive devices to the +17.0 dB figure, we can establish the required output level of each amplifier.

Step 1 - Service from Building 140

The calculations must be made on the highest and lowest frequencies of the passband. That means 5 and 50 MHz for sub-split systems and 5 and 180 MHz for high-split systems. Both cases are addressed in the following calculations:

	5 MHz	50 MHz	180 MHz
Input to Amp. (132)	+17.0	+17.0	+17.0
500 ft. total to Bldg.(140)	+0.55	+1.75	+3.4
DC-12	+0.7	+0.9	+0.95
3-way Splitter	+6.3	+5.7	+5.75
Required output (140)	+24.55	+25.35	+27.1

It appears that a 7 dB pad is required for a sub-split system and a 10 dB pad for a high-split system. Equalizer values can be established by taking the difference between high and low frequencies, i.e. 0 or 1.5 dB for the sub-split system and 3 dB for high-split.

Step 2 - Service from Buildings 128 and 120

All that needs to be done is to add the cable footage between building 140 and 128 to the above numbers:

	5 MHz	50 MHz	180 MHz
Output level (140)	+24.55	+25.35	+27.1
305 ft. (140) to (128)	+0.34	+1.07	+2.07
Required output (128)	+24.89	+26.42	+29.17

Building 120 requires the same return output since there is only a 35 ft. difference in the cable footage.

Pad values are 5.0 dB for a sub-split system and 8 dB for a high-split. Equalizers are 1.5 dB and 3 or 4.5 dB respectively.

Step 3 - Service from Building 148

The total footage in this section is 575 ft. and there is only a DC-12 in the path.

	5 MHz	50 MHz	180 MHz
Input to Amp. (132)	+17.0	+17.0	+17.0
575 ft. total to (148)	+0.63	+2.01	+3.91
DC-12	+12.0	+12.0	+12.0
Required output (148)	+29.63	+31.01	+32.91

The difference here is the 12 dB tap loss of the DC-12 that requires higher levels. Pad values are 1 dB for a sub-split system that originates a 32 dBmV signal and 4 to 5 dB for a high-split system, which has a 37 dBmV output level.

Return Level Calculations (Area 2)
Service from Buildings 158, 152 and 180

Step 1 - Service from Building 158

A +37 dBmV signal at 180 MHz from the building reverse amplifier should arrive at the trunk station input with a level of +17.0 dBmV.

There is a DC-16 and a 4/23 tap in the path. The calculations, in the forward direction, are as follows:

	5 MHz	50 MHz	180 MHz
Trunk input (158)	+17.0	+17.0	+17.0
4/23 Tap insertion	+0.4	+0.5	+0.6
DC-16	+16.0	+16.0	+16.0
Bldg. Amp. output (158)	+33.4	+33.5	+33.6

A sub-split system design needs a 0 dB pad and a 0 dB equalizer. A high-split system design, because of the +37.0 dBmV output, needs a 3 dB pad and a 0 dB equalizer.

Step 2 - Service from Building 152

The distance to building 152 is 545 ft. Since this is the longer distance than the branch to building 180, it is calculated first.

	5 MHz	50 MHz	180 MHz
Input to trunk (158)	+17.0	+17.0	+17.0
4/23 Tap insertion	+0.4	+0.5	+0.6
DC-16 insertion	+0.6	+0.7	+0.75
240 ft. 750-JCA	+0.26	+0.84	+1.63
2-way Splitter	+3.8	+3.7	+3.75
305 ft. 750-JCA	+0.34	+1.07	+3.05
Bldg. Amp. output (152)	+22.4	+23.81	+26.78

A sub-split system would require an 8 dB pad and a 1.5 dB equalizer. A high-split system amplifier needs a 10 to 11 dB pad and a 3 dB equalizer.

Step 3 - Service from Building 180

The only difference for building 180 is the 130 ft. shorter distance. All there needs to be done is to add the differential loss to the above numbers.

	5 MHz	50 MHz	180 MHz
Output Bldg. (152)	+22.4	+23.81	+26.78
130 ft. 750-JCA	+0.14	+0.46	+0.88
Output level Bldg. (180)	+22.54	+24.27	+27.66

Here, again, we need padding for a sub-split system design of 7 to 8 dB with an equalizer of 1.5 dB. The high-split system amplifier needs a 9 dB pad and a 3 dB equalizer. This completes the Area 2 calculations.

Return Level Calculations (Area 3)
Service from Buildings R160, 182, 184 and R170

Step 1 - Service from Building 180

The calculation has already been done for building 158 in Area 2. The building 180 output levels are:

	5 MHz	50 MHz	180 MHz
Bldg. Amp. output (R170)	+33.4	+33.5	+33.6

A sub-split design requires a 0 dB pad and a 0 dB equalizer.
A high-split design needs a 3 dB pad and a 0 dB equalizer.

Step 2 - Service from Buildings 182 and 184

We can combine both buildings in one calculation since the difference in distance is only 10 ft. Do not forget that we used 875-JCA cable in the forward level calculation.

	5 MHz	50 MHz	180 MHz
Input to trunk (R170)	+17.0	+17.0	+17.0
4/23 Tap insertion	+0.4	+0.5	+0.6
DC-16 insertion	+0.6	+0.7	+0.75
550 ft. 875-JCA	+0.5	+1.65	+3.3
3-way Splitter	+6.3	+5.7	+5.75
260 ft. 875-JCA	+0.23	+0.78	+1.56
Bldg. Amp Output (182,184)	+25.00	+26.33	+28.96

Sub-split system: Pad 5 to 6 dB, Equalizer 1.5 dB
High-split system: Pad 8 dB, Equalizer 3 or 4.5 dB

Step 3 - Service from Building R160

There is a difference of 260 ft. minus 170 ft., or 90 ft. All that needs to be done is to add the 90 ft. cable attenuation to the building amplifier output of Step 2.

	5 MHz	50 MHz	180 MHz
Bldg. Amp. output (182)	+25.03	+26.33	+28.96
90 ft. 875-JCA	+0.08	+0.27	+0.54
Bldg. Amp. output (R160)	+25.11	+26.60	+29.5

Since the difference only represents fractions of a dB, the same pad and equalizer values apply as for buildings 182 and 184.

Outside-Plant Design Documentation

This completes the design calculations of any broadband coaxial segment in the outside plant of our hypothetical campus. As mentioned in Chapter 9, the in-building design section, it is recommended to record the design information in an orderly manner together with cable and equipment quantities. The proper collection of all data such as Bill-of-Materials and amplifier set-up forms the basis for a problem-free implementation program.

The Fiber-Optic Outside-Plant Segments

Fiber-Optic Transmission Considerations

Directivity

The transmission of light in a fiber-optic cable is unidirectional. This means that two fiber-optic strands are required for a two-way transmission. At each end of the fiber facility, equipment is required for the transmission or reception of the signals.

Measurement Units

Fiber-optic cables are a relatively new product. Therefore, all distance measurements are expressed in the global standard of the meter. There are 3.287 ft. in a meter, or 1 ft. equals about 30 centimeters. 100 cm equal 1 meter. One thousand meters are a kilometer (km). In a conversion from feet to meters, simply divide by 3.287. 1000 m, or 1 km, represents about 3,287 ft., or a mile represents about 1.606 km.

Modulation Methods and RF Transmission

A transmission on a fiber-optic cable can use the same RF modulation methods as the coaxial cable. An amplitude (AM) modulation signal travels in the assigned frequency band and after conversion of the light to electrical signals, the same RF band continues on the coaxial cable.

Video channels can be stacked in 6 MHz frequency slots (up to 80) and are forwarded to a laser transmitter. They travel as light at RF frequencies through the fiber-optic cable and are reconverted to electrical RF frequencies for further transmission on the coaxial cable.

FM modulated fiber-optic transmitters and receivers also exit and are used for long-distance point-to-point connections and in single-channel configurations, but are high-priced and not required in an HFC architecture.

Analog vs. Digital Transmission

Neither fiber nor coaxial cables care whether the transmission of the signal is in an analog or in a digital format. The HFC broadband infrastructure will support conversion from analog to digital at any time in the future.

Baseband Transmission

At baseband frequencies, i.e. from O to IO MHz, the fiber-optic transmission has an advantage over coaxial cable. The metallic structure of the coaxial cable causes transmission impairments at low frequencies. Equalizers and fine-tuning is required to obtain a flat characteristic of the transmitted signal.

Twisted-pair copper cables have similar problems. But they can handle transmissions at baseband frequencies for short distances using corrective measures in both transmit and receive equipment. This is not to say that broadband video/voice and data can be sent through the local loop of the Telephone Company. The local loop has been an evolutionary process. Many different gauge wires are used and the routing does not resemble a straight line.

Fiber-optic strands do not require special conditioning equipment for baseband since the signal is traveling in the optimum wavelength range of the light frequency.

Wavelength and Attenuation

Broadband signals can consist of voice, data and video channels that are modulated to an RF frequency. This band of frequencies could consist of up to I5O video channels. On the fiber-optic strand, this band of amplitude frequency-modulated carriers must travel at very specific wavelength ranges of the light. Only at these wavelengths is a low attenuation obtained.

By using the wavelength of I3IO nanometer, the attenuation is a mere O.35 dB per kilometer. By using the wavelength of I55O nanometer (nm), the attenuation is even lower - O.25 dB per kilometer.

At I3IO nm, the O.35 dB per kilometer attenuation corresponds to about 48.6 dB at 75O MHz on a coaxial 75O-JCA cable. This means that the fiber-optic strand only has an attenuation of O.72 percent of a coaxial cable.

The low attenuation has promoted the use of fiber-optic cables as the ideal transmission media. It is also obvious from the above that the use of fiber-optic cables is most cost-effective for distances over one-half of a mile. Coupling losses and the cost of transmit and receiving equipment makes the transmission a costly proposition. The last one-half mile in a broadband HFC network is always set aside for coaxial cables.

Star, Tree and Ring

While the coaxial broadband system is a tree-and-branch configuration, fiber-optic strands are using star or ring architectures.

The splicing of fiber must be avoided as much as possible. Each splice produces a transfer problem to the light, which can be expressed in dB. For this reason, the best decision is to use a fiber cable in an end-to-end configuration. Since cable lengths of up to 35,000 ft. can be obtained on one reel; a IO km distance can be made without the need of splicing.

An HFC system then is basically a fiber star or ring with a coaxial tree-and-branch distribution area connected to the end of the fiber. The translation location between fiber and coaxial cables is usually referred to as the fiber-node.

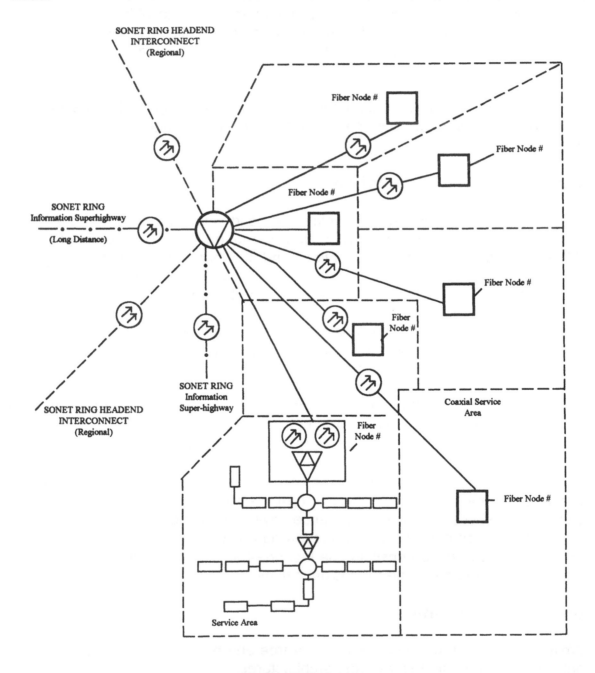

Fig. 10-4 Typical HFC Architecture

Fig. 10-4 shows the typical architecture of an HFC system in the outside world. The location in the center is the fiber hub, headend, gateway or processing center for voice, data and video. From the processing center, the fiber-optic cables are emanating in a star architecture. Each fiber-node location is the starting location for a coaxial network using a tree-and-branch architecture. Broadband interconnections to other processing centers are in a ring architecture so that all information is protected by alternate routing.

The regional interconnect may be AM modulation on fiber in an analog format or SONET rings with ATM switching equipment for digital transmission of voice, data and video. The same applies to long-distance traffic to and from the processing center. In our hypothetical campus layout, Fig. 7-3, the processing center is the headend in building 132, which is the campus connection to the outside broadband world.

The Forward Transmission Design

Cable Selection

There is only one type of fiber strand that qualifies for the transmission of broadband voice, data and video channels. It is the single-mode fiber that operates on a wavelength of 1,310 nm.

What we are mainly interested in is a fiber cable that has a good tensile strength. The types of available single-mode fiber cables consist of feeder cables with armor, central tube cables with or without armor and loose-tube cables with or without armor. There is no electrical difference between any of these cables. The only major difference is the number of fiber strands and the filling compound between the bundles. 12 fiber strands are contained in a bundle and are identifiable by standard industry color coding.

Armored feeder cable is available with 2 to 12 fiber strands.
Central tube cable is available in 2 to 48 strands for nonarmored and for 2 to 96 strands in the armored version. Loose-tube cable is for large strand-count requirements, i.e. from 2 to 216 fiber strands in nonarmored and armored version.

A big handling difference is the cable weight per 1000 ft., which ranges from 57 lbs. per 1000 ft. for the armored 12-strand feeder cable to 275 lbs. per 1000 ft. for a 216-strand armored loose-tube cable.

The electrical characteristics for all types are as follows:

Wavelength	1310 nm	1.550 nm
Attenuation (dB/km)	0.35	0.25

In our hypothetical campus layout, we have a total of 5 fiber-node locations. We need at least one fiber for forward and one fiber for return. A 12-strand fiber cable between the headend and each fiber-node location appears to be a good choice that would safeguard any future requirements.

Based on this assumption, the mechanical specifications of nonarmored or armored central tube cable appear attractive.

	nonarmored	armored
Outer diameter (inch)	0.49	0.49
Weight per 1000 ft (lbs.)	105	115
Max. tensile strength (lbs.)		
during installation	600	600
Crush resistance (lbs.)	500	1,000

Only you, with the intimate knowledge that you have of the campus, can make the decision to use armored or nonarmored cable.

Cable-Routing Options

It is important to determine the optimum routing of all fiber cables.

From a review of the campus layout of Fig. 7-3, the following cable connections are necessary:

	No. of Strands	Footage with spoilage
Building 132 to R100	12	1,200
Building 132 to R110	12	1,700
Building 132 to R150	12	1,600
Building 132 to R170	12	2,400
Building 132 to 158	12	1,000

The footages include about 10% of the distance for spoilage.

We discussed in the foregoing that high-speed data, voice and video may be combined in a digital format. We also discussed that a high-speed survivable ring architecture provides for uninterruptable communication. Maybe the costs of SONET multiplexing equipment and ATM switches will drop in the future to make such an installation affordable. Since the implementation plan for the campus HFC system is still in the conceptual stage, it may be the right time to complete the fiber infrastructure and to permit the establishment of digital traffic at 2 GHz.

The bottleneck in this plan is the new conduit section of 210 ft. Assuming that this conduit will be installed, additional fiber cables could be pulled simultaneous to the implementation of the HFC system.

This optional fiber-cable routing plan would establish the availability of one fiber cable in each direction at the building entry MDF of each building.

          I48 to MH-A to I32
          I48 to MH-A to MH-AI to I58
          I58 to MH-AI to I52
          I52 to MH-AI to I80
          I80 to I82
          I82 to MH-E to RI60
          RI60 to MH-E to I84
          I84 to MH-E to MH-O to RI70
          RI00 to MH-B to MH-BI to RII0
          RII0 to MH-BI to RI50
          RI50 to MH-BI to RI70

The ring is complete, except it does not cover building I28, I40 and I20. Since the service to these buildings is via steam tunnel, it would be easy to install the fiber in the same raceway with the coaxial cable.

The routing would be:

          I32 to I28
          I28 to I40
          I40 to I20
          I20 to I32

We have added a second cable to the HFC infrastructure. The installation costs of pulling two cables vs. one cable are no more than a surcharge and can be a wise investment for the future.

The Transmission Equipment

The Transmitter

The difference between fiber-optic I3I0 nm transmitters is the type of the light-emitting laser. The choice is either the lower powered Fabry Perot laser or the higher powered distributed feed-back (DFB) laser.

In the hypothetical campus system the distances are short, but we need to transmit many channels. Transmitters are available for sub-split and high-split systems with an upper frequency of 200 MHz, 550 MHz and 750 MHz.

The Link Budget

The higher the number of channels, the greater the need for high power. For instance, a 24-channel Fabry Perot transmitter has an output power level of about 20.5 dBμW per channel, and since the receiver requires a receive power level of around 13.0 dBμW, the difference between the two is referred to as the link budget, as in our example 9.5 dB.

A 60-channel DFB transmitter, on the other hand, has an output power level of 24.5 dBμW per channel and the same receive power level requirement of 13.0 dBμW. Therefore, the link budget is 11.5 dB.

Some important specifications of optical transmitters are shown below:

	Fabry Perot	DFB low power	DFB high power
No. of channels	24	60	60
Transmit power (dBμW/Ch}	+23.5	+22.5	+24.5
RF Input level (dBmV)	+32.0	+32.0	+32.0
Receiver input (dBμW/Ch)	+13.0	+13.0	+13.0
Composite triple beat (dB)	56.5	56.5	56.5
Optical budget (dB)	10.5	9.5	11.5

Why do we worry about link budgets, if the attenuation of the fiber-optic cable is 0.35 dB per kilometer?

It is correct that our campus layout does not require that light to travel very far. The longest distance is between the network center at building 132 and R170, which was calculated to be about 2,400 ft. This is about 730 m, or not even 1 km.

Optical Couplers

At the network center, the location of our future transmitter, there are 5 fiber-optic cables for the 5 fiber-nodes. Somehow, we have to get the optical power into all 5 cables. This is done by using fiber-optic couplers, which have about the same insertion losses as the coaxial splitter.

Again, there are even and uneven 2-way couplers available as well as 4-way devices. Couplers are usually packaged in a stainless steel, 3 to 4-inch long housing with a diameter of less than one-tenth of an inch.

The following electrical data is typical for fiber-optic couplers:

	Coupling Ratio	Primary Port	Insertion Loss (dB) Secondary Port(s)
2-way Coupler	50-50	3.7	3.7
2-way Coupler	30-70	2.0	6.0
2-way Coupler	20-80	1.3	7.8
4-way Coupler	50-50	7.4	7.4
4-way Coupler	40-60	6.2	9.3

It is concluded that the coupler insertion loss consumes most of the link budget.

For the 5 fibers that are coupled to the headend transmitter, it is important that the couplers are selected in a manner that will provide approximately equal output power to all fibers.

Receivers

Receivers convert the optical frequencies back to RF frequencies. Essentially, all channels fed to the transmitter input will be available at the output of the receiver. Receivers are available for both sub-split and high-split systems and feature a low and a high RF output level.

The low RF output level version has an output of +15.0 dBmV, which is sufficient for signal processing at the headend and the outlying receive locations.

Typical specifications are provided below:

Upper frequency	200 MHz	550 MHz
Optical input (dBµW/Ch)	+13.0	+13.0
Carrier-to-noise ratio (dB)	53.5	51.5
Compst. triple beat (dB)	122.0(24)	108.0(60)
Receive wavelength (nm)	1,310±20	1,310±20
Channel capacity	24 and 32	60 and 80
RF output level (dBmV)	+13.0	+13.0

All fiber-optic equipment manufacturers offer receivers in standard cable housings for outdoor installation. Rack-mountable equipment is recommended, which can be mounted in a standard 19-inch wall panel placed on the wallboard.

Forward Transmission Calculations

The Trial Design

All that needs to be done is the determination of the most efficient coupler combination and the cable attenuation for the longest run. The result will lead to

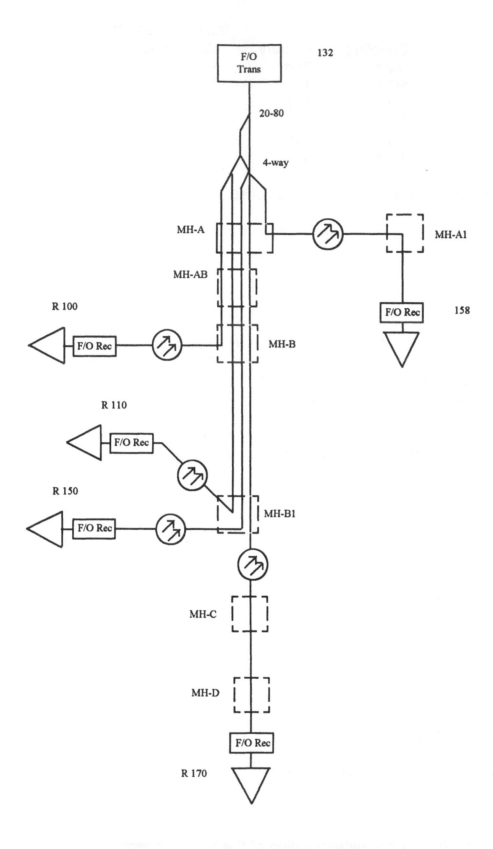

Fig. 10-5 The Forward Fiber Star

the selection of the required transmitter. Fig. IO-5 shows the routing of the fiber cables to the 5 node locations. Starting with the receiver's optical input level of +I3.O dBµW/channel, we can add the losses with a plus sign to obtain the required transmitter output level:

Receiver input (dBµW)	+I3.O
2,4OO ft. SM fiber (dB)	+O.26
2-way Coupler 2O-8O	+7.8
Req. Xmtr outp. level (dBµW)	+2I.O6

For 6O-channel loading, the low-power DFB transmitter is available with an output power level of +22.5 dBµW/Ch. The trial design has been successful.

Final Design

The final design can be conducted in the forward direction for all five locations using a 4-way coupler connected to the low-attenuation output of the 2O-8O coupler and determine all five receiver input levels.

RI7O Receiver Level:

XMTR Ouput Level (dBµW/ch)	+22.5
2O-8O, 2-way Coupler (dB)	-7.8
2,4OO ft. SM Fiber (dB)	-O.26
Receiver Input Level (dBµW/ch)	+I4.44

RI5O Receiver Level:

XMTR Output Level (dBµW/ch)	+22.5
2O-8O, 2-way Coupler (dB)	-I.3
4-way Coupler (dB)	-7.4
I,6OO ft. SM Fiber (dB)	-O.I7
Receiver Input Level (dBµW/ch)	+I3.63

RIIO Receiver Level:

XMTR Output Level (dBµW/ch)	+22.5
2O-8O, 2-way Coupler (dB)	-I.3
4-way Coupler (dB)	-7.4
I,7OO ft. SM Fiber (dB)	-O.I8
Receiver Input Level (dBµW/ch)	+I3.62

RIOO Receiver Level:

XMTR Ouput Level (dBµW/ch)	+22.5
2O-8O, 2-way Coupler (dB)	-I.3
4-way Coupler (dB)	-7.4

```
        I,2OO ft. SM Fiber (dB)           -O.I3
        Receiver Input Level (dBµW/ch)  +I3.67
```

I58 Receiver Level:

```
        XMTR Output Level (dBµW/ch)    +22.5
        2O-8O, 2-way Coupler (dB)      -I.3
        4-way Coupler (dB)             -7.4
        I,OOO ft. SM Fiber (dB)        -O.II
        Receiver Input Level (dBµW/ch)  +I3.69
```

The forward design is complete. It is noted that optical receivers can receive higher levels than +I3.O dBµW/ch. Every I dB of optical level change will cause a 2 dB change in the RF output of the receiver. RF attenuators are available for the receiver to provide the +I3.O dBmV output level minimum requirement.

The Return Transmission Design

Transmission Equipment

Transmitters and receivers for the return direction are identical to the equipment discussed in the Forward Transmission Design. There is one difference, however, and that is the channel loading.

Since we used a high-split architecture in the coaxial segments of the design, only a total number of 3O simultaneous transmissions can be made from each of the five fiber-node locations.

It is, however, very unlikely that 3O simultaneous transmissions are originating in a single building, even in the case of RIOO and RIIO with I92 broadband outlets. For instance, the use of 8 or I6-channel transmitters could save over $IO,OOO per transmitter location.

The specifications below show the differences between the transmitters:

	Fabry Perot	Fabry Perot	DFB Low Power
No. of Channels	4	8	32
Transmit Power (dBµW/ch)	+25.O	+23.O	+22.O
RF Input Level (dBmV)	+32.O	+32.O	+32.O
Receiver Input	+I3.O	+I3.O	+I3.O
Comp. Triple Beat (dB)	62.5	62.5	62.5
Optical Budget (dB)	+I2.O	+IO.O	+9.O

In the return direction, we do not have to worry about the link budget. Each fiber-node transmitter is connected to the receiver in building I32 without any coupling devices.

The receiver units are identical to the 200 MHz model described in the Forward Design.

Return Transmission Calculations

With a I0.0 dB link budget for an 8-channel transmitter, and using the cable attenuation values from the forward design calculation, do not bring the receiver level to +I3.0 dBμW/ch. The cable attenuation numbers vary between -0.II and -0.26 dB. Subtracting these numbers from the transmitter ouput power level of +23.0 dBμW/ch, we can expect receiver input levels in the range of between 22.74 to 22.89 dBμW/ch.

A I0 dB attenuation is required in each receiver to keep the RF level within the range of about +I3.0 dBmV.

Alternative Return Transmission Architectures and Cost Budgets

The geographics of the hypothetical campus area and, for that matter, of any large multi-building area, permits three different return transmission architectures. Each of these architectures deliver high-quality broadband channels to the network or operations center. The difference lies in

  a) the quantity of channels available
     from each building and at the same time
and
  b) the capital outlay for equipment

A comparison of these alternatives is presented below. All three have been based on the assumption that each building must accommodate four simultaneous transmissions.

Alternative I - The Fiber-optic Return in a Star Topology

The return transmission calculations of the last section are representative of this approach. A fiber-optic transmitter is located at each fiber-node location for optical transmission in a direct line to the network operations center. Fig. I0-6 illustrates the fiber-optic return system.

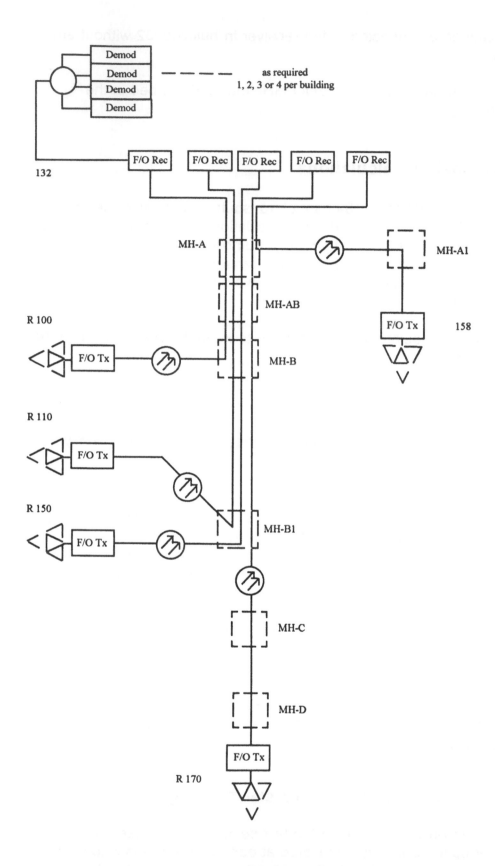

Fig. 10-6 The Return Fiber Star

In this architecture, each of the buildings can return a substantial number of broadband channels. However the return requirement has been based again on the 4-channel-per-building requirement.

- At R170, a 16-channel transmitter can accommodate four simultaneous return transmissions from R170, R160 , 182 and 184

- At R100, R110 and R150, 4-channel transmitters can provide four simultaneous transmissions from each building

- At 158, a 16-channel transmitter can provide for five simultaneous transmissions from buildings 158, 152 and 180 (only 4 per building are used)

- The return transmission for buildings 148, 120, 140 and 128 is coaxial. 28 return channels can be proportionately allocated for each of the buildings (7 each). However, for this comparison, only 4 per building are used

This topology then provides a total of 59 return-channel allocations. All signals arrive in a RF format at the output of the fiber-optic receivers and require demodulators for conversion to baseband and further processing through a matrix switch. Each return channel becomes an input in this video/audio or video/dual-audio routing switch.

The equipment complement, therefore, consists of
3   4-channel FO transmitters
2   16-channel FO transmitters
5   FO receivers
56   RF demodulators (full occupancy)

The price tag for fiber-optic equipment is coming down, but still presents a substantial burden. It is estimated that the cost of the above equipment complement is in the range of $60,000 to $80,000.

Alternative 2 - The Coaxial Return
in a Tree-and-Branch Topology

The coaxial return system has already been calculated for buildings 182, R160, 184 to R170 (Area 3), from 120, 140, 128 and 148 to 132 (Area 1), and from 152 and 180 to 158 (Area 2). The buildings not presently served by a coaxial return path are R170, R100, R110, R150 and 158. (Fig. 10-7)

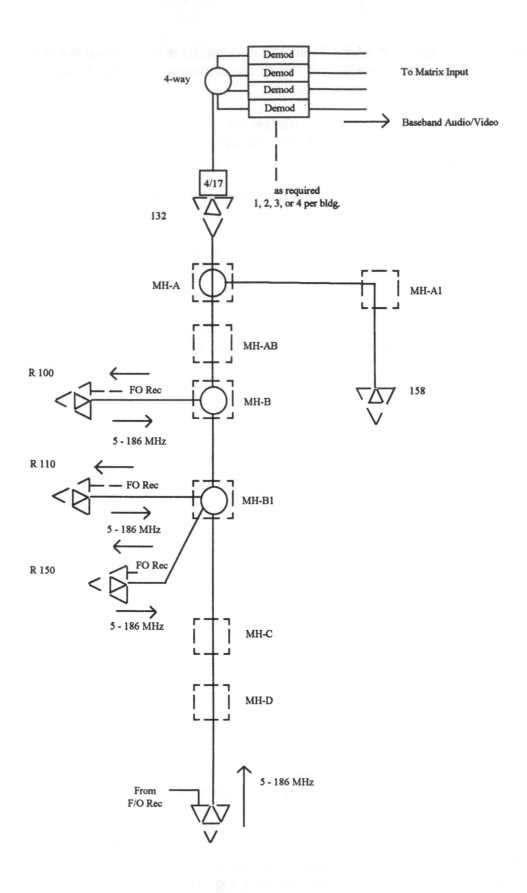

Fig. 10-7 The Coaxial Return

Since the upper frequency of the return is at 186 MHz, a passive 750-JCA cable with splitters in MH-B and MH-BI can be interconnected with 132 without the need for any amplifier. The return cable would simply be connected to the return output of each building amplifier. The same cable return would interconnect buildings 152, 180 and 158. Fig. 10-7 illustrates the single-cable coaxial return system.

- 4-channel return per building can be provided for buildings 182, R160, 184, R170, R150, R110 and R100 by pulling a 750-JCA coaxial cable between R170 and 132 with spurs to R150, R110 and R100 from splitters in MH-B and MH-BI.
  The total number of simultaneous return channels is 28 (4x7), which is just the right number for a 5-186 MHz transmission system.

- 4-channel return per building can be provided for buildings 152, 180 and 158 by pulling a 750-JCA coaxial cable between building 158 and 132 (Area 2).
  The total number of return channels in this example is 12, but can be expanded to 28 channels.

- 4-channel return for buildings 120, 140, 128 and 148 already exists (Area 1).

This elegant method does not require any transmission equipment or conversion from electrical signals to light and back. Each building has 4 channels reserved for simultaneous transmissions, for a total of 56 broadband transmission channels. The only additional equipment required are the 56 demodulators (full occupancy), which can be procured in a cost range of $36,000 to $42,000.

Alternative 3 - The Fiber-optic Return in a bi-directional Ring Topology

The possibility of using fiber-optic cables in a ring topology for future SONET services has been discussed in the section Cable Routing Options. Assuming that the 200 ft. of new conduit has been installed, all buildings can be interconnected with 12, or more pair, single-mode fiber cables. The fiber-optic splices in each building will absorb the link budget quickly and a 2-way coupler is required to send optical transmission in two directions as well as accept optical reception from either direction. Even though, the distances between the building are short, it is important to design such a system with care and determine coupler and splicing losses accurately. Fig. 10-8 provides a routing diagram for a uni-directional fiber ring. Route redundancy can be obtained by duplicating the fiber count.

Assuming, again, the requirement for four simultaneous return transmissions from each building, we can develop the following list of equipment:

14   4-channel FO transmitters
56   FO receivers
56   RF demodulators

Couplers and splicing has not been considered in this list since the resulting redundancy would not permit an equal comparison between the alternatives.

The price range of this equipment is estimated between $150,000 to $180,000 and higher than either Alternative 1 or 2.

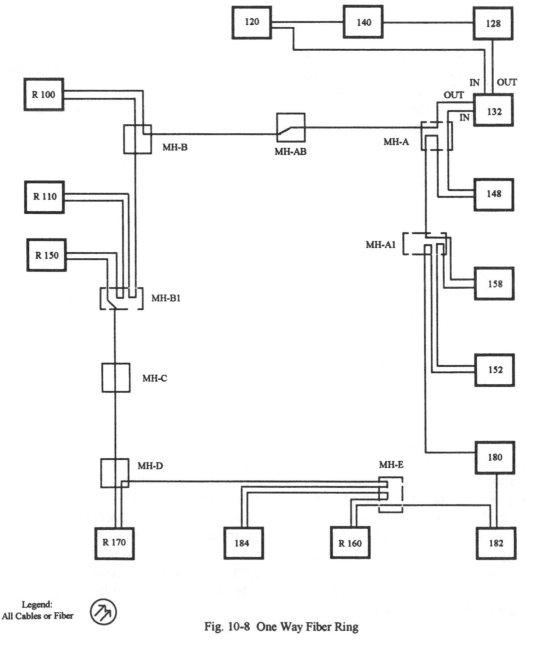

Fig. 10-8 One Way Fiber Ring

Conclusions

While the forward HFC system design was based on increasing the quality of service and reducing maintenance costs, the return design does not benefit from the use of a fiber-optic return transmission. The most economical architecture is the coaxial tree-and-branch.

If, however, the decision has been made to use fiber-optic cables as much as possible, and if the funding is available, the fiber-optic return in the star architecture is the winning alternative.

In the case, however, that your plans call for a SONET ring for high-speed data expansion, the ring topology of Alternative 3 is the logical choice. As was pointed out, a ring topology of the fiber-optic cables can provide survivable communication throughout the campus.

The routing of the broadband HFC network coincides with that of a high-speed data network, an advantage for maintenance and network management. The Alternate 3 example, of course, can be expanded for forwarding 60 channels to each building. Instead of 5 optical receivers, a total of 14, or one for each building, are needed.

It is noted that the step to bi-directional return traffic is an expensive one since a total of 112 fiber-optic receivers and demodulators would be required.
It is also noted that, if you proceed with the ring architecture of Alternative 3, no coaxial broadband outside plant would be required. All the forward and return outside-plant design calculations for Areas 1, 2 and 3 can be discarded.

The decision is yours and directly related to political and economical factors in your enterprise.

If your budget contains an extra $100K plus, or if the additional fiber-cable footage can be justified on the basis of a high-speed data backbone, then the ring topology of Alternate 3 is the preferred architecture. This decision in favor of the fiber ring would not be made to achieve a better performance, but to meet any kind of future data or HFC broadband requirements and to achieve the highest degree of flexibility in circuit assignment and utlization of the spectrum.

If, on the other hand, the funding is low and you are looking for the highest return on your investment, the coaxial tree-and-branch return architecture may serve your needs for the next decade or two.

In case that it is your preference to have identical cable routes for forward and return transmissions in your system, then the star return architecture of Alternate 1 may be your choice.

Chapter 11

HFC Installation Considerations

Now, that you have completed the design of your HFC broadband network, it is time to formulate your plans for implementation. If it is your plan to issue an IFB, Information for Bidders, or an RFP, Request for Proposal, it is important to cover the installation particulars in considerable detail.

The more specific you can be in telling the bidder what you expect, the more competitive the bidding process will be.

By specifying routes and installation methods in great detail, you will encourage the more seasoned contractor and discourage the inexperienced. Your HFC network has to coexist with existing wires and cables for a long time. As a matter of fact, your HFC network may become the only surviving infrastructure. It may become the backbone for telephony expansion and for a new high-speed data concept. It may carry PCS traffic and HDTV video-on-demand. An HFC system is not subject to moves and changes. Every outlet offers identical service capabilities.

It is, therefore, of primary importance that you plan the installation of the HFC network in detail and develop installation standards that cover all eventualities.

Your new HFC network may be the last infrastructure that you will have to implement. A detailed treatment of routing and installation directives will give you some assurance that you are getting your money's worth and that there will not be too many "extra scope" requests by the contractor. From your past experience, you know that it is imposssible to write a set of installation instructions that fully protects you against claims by the contractor for out-of-scope efforts. It is not required that your specifications are perfect, but it will save money and reduce your involvement in the supervision of the contractor if you make an effort to strive for perfection.

Planning for the Installation

Armed with the Design Checklist from Chapter 7, The Equipment Specifications of Chapter 8 and your System Design Calculations of Chapters 9 and IO, it is appropriate to develop an Installation Plan.

When contemplating such a plan, asked yourself two questions:

- where do I want the cable and equipment installed?

- how do I want it installed?

It is reality that most buildings or areas will have cabling installations that were undertaken by several different people who have been transferred or by outside contractors. Wiring, in the past, was probably installed by each department on an as-needed basis with little or no coordination, overall planning or documentation.

Wiring systems grow as a function of ever-increasing connectivity needs. It is estimated that wiring collections in some well-known companies are only 20% utilized. It is often much easier and less expensive to install another cable than to find the unused one, or to remove obsolete wiring.

Not so with the HFC network. Since it consists of single-mode fiber-optic cables and large diameter coaxial cables that will not have to be changed to accommodate future service requirements, the cable routing demands respect and special considerations.

Existing Plant Inventory

To locate congested areas and to assess the difficulties of routing your proposed HFC system, it is recommended to first collect any documentation that may be available of the existing plant.

If you are a large enterprise, your facilities department may have just transferred all existing cable and wiring information from blueprints to CAD drawings. If that is so, then all you have to worry about is whether the information is accurate and updated.

If you are, however, an enterprise of a smaller size without a facility department and without any data-base information, your task just got a bit bigger. It is not the purpose of this chapter to insist on an accurate inventory of all existing communications facilities. If there is a good record system, it will assist you with your planning. But, if there is no inventory or documentation of existing facilities, how can you determine where and how to install the HFC facility?

Since some information is better than none, it is recommended to collect all outside-plant blueprints and inside-plant floor plans that you can find. These drawings will give you some orientation and provide you with baseline information.

What you want to accomplish is the determination of the route and method of installation of the new HFC cabling within the spaces of the existing communications wiring. There is only one way to accomplish this and that is a physical survey.

But, before making such an extended field trip to determine where the HFC cable should go, it is even more important to determine all installation standards.

The rationale here is that it will not be helpful to walk through a steam tunnel or climb into a riser shaft without having a picture of the installed cable in your mind. To form this mental picture of the HFC system in place, installation standards have to be determined first.

The Universal Wiring Plan

The universal wiring plan addresses the handling of copper, fiber and coaxial cables. Chances are that somebody before you worried about IBM mainframe cabling and wrote a universal wiring plan. If that is so, review it, retain all references to UL and NEC and replace the rest with your own installation standards.

This remark may not be appropriate in most cases, but it is made to emphasize that an HFC system is a different architecture than type 3 or type 5 wire bundles. While the in-building wiring of the HFC system may require new installation standards, the outside-plant section of your universal wiring plan may be usable as is. Only your careful review can pave the way for the correct decision.

Installation Standards of the HFC Network

New UG Duct Installation

New conduit routing requires a detailed engineering survey of the route. Distances of more than 300 ft. require the placement of manholes.

The sizing of manholes is an important consideration when planning for the future. A six-by-six feet manhole is usually sufficient to accommodate even the most difficult cable-pulling task.

Electrical wires and telecommunication wires must be separated and shall have individual conduit routes.

A dual 4-inch conduit system is recommended as a minimum. Triple or quad inner-ducts, equipped with pull ropes, are recommended to offer separation between fiber-optic and coaxial cables.

Inner-ducts shall be composed of extruded, corrugated wall, coilable PVC tubing conforming to the following minimum standards:

- A-12353 cell classification as defined in ASTM D-1784, or A-11331 cell classification as defined in ASTM D-4396

- Tensile properties, external loading and impact resistance shall conform to the standards as defined in ASTM D-638, D-2412 and D-244; and NEMA TC-2 specifications for electrical plastic tubing

Ovalization of the inner-duct on the reel shall not exceed 5%. The reel set of the inner-duct shall be removable without buckling or kinking. All inner-duct shall be free of holes, splits, cracks, blisters, inclusions and performance-affecting imperfections.

The new conduit route must be carefully measured and staked before commencement of construction. A detailed drawing showing the proposed route with measurements and gradelines shall be prepared to aid the contractor in the implementation in accordance with your plan.

LB junctions, commonly used for power wires, must be avoided. All rigid steel conduit runs shall be equipped with sweeping 45, 60 or 90-degree sections, as required.

While determining the route for the new conduit, it is recommended that cable locators be used to determine any existing buried cabling. If existing cables must be crossed, it is required that the depth of the cables be accurately determined and that a drawing be prepared containing the details of the proposed crossing.

Installation in Steam Tunnels and Buildings

Steam tunnels, whether active or abandoned, are ideal pathways for telecommunication cabling. It is, however, recommended that HFC network cabling not be lashed to existing cable brackets, but run in conduit.

As discussed before, a dual 4-inch conduit bank with triple or quad inner-duct with pull ropes is recommended to provide a secure raceway for present or future installations.

The route of the conduits has to be engineered in great detail. Steam tunnels do not always offer straight interconnections for long distances. The location and type of mounting brackets must be defined in addition to sweeps and breakout locations.

LB junctions shall not be used. All sweeps, whether 45, 60 or 90 degrees, shall have a radius twice as large as the minimum bending radius of the largest cable.

In both steam tunnel and in-building basement routes, it is important to add the number of degrees between pull boxes. Rotations of more than 360 degrees shall be avoided if possible. Junction or pull boxes shall be two-by-two or four-by-four feet and strategically placed in a manner that will allow convenient cable setup.

The preparation of a drawing itemizing the conduit route, supports, sweeps and junction boxes is recommended.

Cable Installation in UG Ducts

All fiber cable must be installed in accordance with manufacturer-recommended tensile and bending specifications. In no case shall the pulling load exceed 600 pounds in a straight pull or 280 pounds in a pull having a 90-degree bend.

Ideally, there shall be no more than two 90-degree changes of direction in any single pull. If there are more than two 9O-degree changes of direction, however, back-feeding or center-pulling techniques must be used. No residual tension shall remain on the cable after pulling.

The manufacturer's minimum bend radius specifications shall not be exceeded. In no case shall the bend radius be less than 16 times the cable's diameter.

Lubrication shall be used when installing fiber and coaxial cables. This lubricant must be manufacturer-guaranteed to be nondestructive to the cable sheath or any portion of the inner-duct.

The following installation requirements shall be followed when installing fiber-optic or coaxial cables:

- A nylon pull rope must be used for the entire distance of the pull

- Corner roller, pulleys and suitable cable guides must be utilized in manholes and at all corners to prevent the cable from touching conduit entrances and to avoid exceeding the minimum bending radii

- A Kellums grip will be used over the capped and taped cable end. A power winch with tension-control device shall be utilized when pulling the cable

- In manholes or steam tunnels where future splices and passive devices are to be installed, the contractor shall provide enough cable tail in order to lay the cable neatly inside the cable racks to permit TDR testing before splicing

Equipment-Mounting and Cable-Splicing

No splicing of fiber-optic cables is permitted. Fiber-optic cables can be spliced to pigtails within the fiber storage and termination panels.

Wallboard-mounting is considered standard for coaxial equipment at MDF or IDF locations. The sizing of the O.75-inch plywood panel is depending upon the equipment required. Plywood panels made from noncombustible plywood are to be used. A two-by-two feet panel is considered the minimum size.

The equipment layout shall be determined by the contractor unless specific drawings are provided. All contractor-created equipment-mounting arrangements must be submitted for approval before commencement of any activities.

Fiber-optic termination units, F/O transmitters and receivers can be mounted on a 19-inch standard-size wallmount bracket. Coaxial devices equipped with strand-mounting clamps can be mounted on the wallboard using suitable angle brackets made from flat or round aluminum.

When pulling the cables to the MDF locations, the cable tails shall not be less than six feet or longer than fifteen feet in length and shall be coiled. The cable ends shall be lashed or taped together and fastened with tie wraps loosely to the plywood board without leaving any stress on the cable. After completion of TDR and OTDR testing, the cable is ready for splicing.

Splitters and directional couplers in manholes shall be mounted to cable trays or on cable support brackets.

All coaxial cable connectors in the outside-plant segment shall be secured with heat-shrinkable tubing and coated with water-proofing compound.

To assure correct splicing and connectorization of all coaxial cable and equipment, the contractor shall provide his cable-termination specifications and supply a list of coring tools that will be used on the project for approval and before commencement of splicing activities.

Heat-shrink tubing and weather-proofing methods are not required at in-building MDF or IFD locations.

All fiber, coaxial and power-supply cables shall be dressed and tie-wrapped to provide for an orderly and neat appearance. Coaxial equipment such as amplifiers, passive devices and multitaps at MDF and IDF locations shall be securely mounted to the wallboards and interconnected using housing-to-housing connectors.

All amplifier housings at building entrance points or at BDF locations shall be grounded to a common building ground location using a solid copper ground wire of no less than No. 8 gauge. Utilizing suitable heavy ground lugs to assure positive grounding of the amplifier housing will assure the protection from cable sheath currents entering the ISP segment of the distribution plant.

Grounding shall be provided for the protection of personnel and equipment conforming to all applicable codes and standards including, but not limited to, the current National Electric Code (NEC), articles 250 "Grounding" and 800 "Communications Circuits", and the current versions of the National Fire Protection Association (NFPA), publications NFPA 70E "Electrical Safety Requirements for Employee Workplaces", NFPA 75 "Protection of Electronic Computer/Data Processing Equipment" and NFPA 78 "Lightning Protection Code".

These are minimum requirements and do not replace state, local or other applicable codes, laws or regulations, which may have priority. The vendor is required to determine, research and adhere to all pertinent requirements in this regard.

Installation in Risers

Horizontal and vertical risers consist in most cases of a single coaxial cable. Only riser-rated cables are to be used. If coring of floors is required, the penetrations must be treated with approved fire-stopping material, conforming to UL listing #1479.

Riser cables shall be routed without violating the cable's minimum bending radius. Vertical runs shall be supported with suitable clamps every six feet.

During cable installation, coaxial cable tails of not less than 6 feet shall be coiled, taped and loosely connected to the wallboard until splicing commences.

Horizontal risers shall be routed over the ceiling and connected to supporting members every 3 to 4 feet. The cable shall be lightly tensioned to prevent a sag in excess of 2 inches. The use of tie-wraps to secure the cable to supporting members is permitted.

Horizontal riser cable will be riser-rated or plenum, if required. Passive equipment and multitaps, required at the end of the horizontal riser, shall be mounted securely to existing structural supports. Special mounting hardware will be required since the use of tie-wraps for the securing of equipment is not permitted.

The Installation of Service Drops

Riser-rated or plenum type 6-series supershielded cables shall be used exclusively. All service drops are 150 feet long, unless system design requires a different footage standard.

Short service drops require the excess cable to be coiled and stored. Do not exceed the minimum bending radius specified when coiling excess cable.

Installation using Wire Trays

When a substantial number of service drops are routed in the same direction, i.e. in the ceiling space above a hallway, the use of wire trays is recommended. A wire tray can accommodate type-5 data and type-3 voice wiring at the same time and provide a solid support. Service drop cables shall be bundled, tie-wrapped every four feet and orderly located in a section of the wire tray. Excess footage can be coiled at the point of departure from the wire tray and before feeding the drop within the wall to the outlet location. Excess cable shall be neatly coiled, tie-wrapped and fastened to the wire tray or nearby supporting members.

276

Over-the-Ceiling Installation

Single drops or bundles of not more than IO drops can be routed over the ceiling, bundled and tie-wrapped every 3 feet. In addition, tie-wraps shall be used every 3-4 feet to fasten the bundle, or a single drop to an existing structural member.

The coiling of excess drop cable, again, can be provided at the feed point to the outlet. Excess cable shall be neatly coiled, tie-wrapped and fastened to a structural member without exceeding the minimum bending radius specifications.

Installation in Molding

Older building construction without false ceilings may require the use of molding for the distribution of service drops.

The size of the molding is determined by the maximum number of service drops feeding in a particular direction when leaving the IDF location. Identical-size molding shall be used everywhere and where required, even for the longest single drop.

The routing of the molding and the type proposed for use shall be submitted for approval by the owner before commencement of the installation.

Excess service drop cable, in this case, can only be coiled at the IDF location. It is recommended that a separate wallboard be used in the IDF to store the excess drop cables. Each drop-cable excess must be coiled and harnessed separately and then bundled together.

Service Drop Terminations at the IDF

All service drops terminate at the IDF location at multitap ports. The connection is made using F-connectors. Unused multitap ports must be terminated with F 75-ohm male terminators.

Service drops shall be bundled per multitap using tie-wraps. The bundles must be dressed neatly to form an orderly and aesthetic appearance with sufficient slack to minimize tension on the F-connectors. Bundles from each multitap shall be harnessed together in an orderly manner when leaving the IDF for distribution to the outlets.

Universal Outlets

Modular faceplates are available to combine voice, data and broadband connections. F-female bulkhead connectors are required for broadband HFC outlets. Bi-directional service requires one port for receive and one port for transmit service.

The makeup of the universal outlet depends upon the wiring to be implemented. It is the contractor's responsibility to propose a standard faceplate layout and secure approval by the owner before commencement of installation.

Cable Marking

All cables shall be labeled within 6 inches of either cable end. The cable numbering system must be submitted to the owner for approval before attaching final labels.

Labels shall be of waterproof material using a waterproof adhesive attachment. All numbers shall be printed and indelible. The contractor must coordinate with the owner to ensure the compatibility of the labeling system with the existing facility management systems.

Distribution cables may carry building or MDF numbers. Service drop cables may carry floor and room numbers for easy identification of the cable routes.

Finalizing the Installation Plan

Now, that you have developed the installation standards for your HFC network, it is time to

-     see if what has been written, can be applied

-     determine routing details

-     determine any exceptions to the rule

-     make any supplementary sketches of problem areas

The time is right for a final detailed physical review of the HFC cable route.

Outside-Plant Routing

In steam tunnels, the verification of the conduit route may include:

-     locating the HFC conduits away from existing loose cable

-     determining the location of existing power wires, if any, and maximizing the separation of the proposed HFC system.

- determining special attachment hardware

- determining the exact location of every sweep or directional change

- making a sketch of the proposed conduit route with construction details

- verifying all distances one more time

In existing conduit sections, you have already compiled a detailed usage plan when you prepared the design checklist. You only have to open the manholes again if you want to designate a conduit for the HFC system. In our hypothetical campus layout, we located at least one empty 4-inch conduit. It may be a good idea to specify the installation of inner-duct with pull rope as the first task of your implementation program. Doing this will save 2 inner ducts for future cabling requirements.

Coaxial outside-plant segments may require the installation of splitters in some manholes. The following is recommended:

- determine where the splitter is to be installed

and
- how it is to be mounted

If you do not stipulate these steps, you may find that the contractor will install the splitter at a convenient location to him and it will be used later as a foot rest by everyone descending into the manhole.

Inside-Plant Routing

It is a good practice to revisit all MDF building-entry locations, riser spaces and IDF rooms to confirm that your installation plan can work. When you visit these locations, asked the following questions:

- is there enough space for the wallboard?

- is there a IIO Vac power outlet?

- is there any UPS power and where?
  (the provision of electrical power is usually
  the responsibility of the owner)

- how are the cables to be routed between
  building-entry point and wallboard?

- what is the routing of the service drops within and out of the MDF?

- how would I like to see the equipment placed on the board?

- where do I want to see the riser cable installed? (the old copper wires are blocking my view)

- are there empty conduit sections interlinking the floors?

- is any coring to be done?

- what about fire stops? (none of the existing cabling has any)

- where are any convenient structural members to connect the drop wires to in the ceiling space over the hallway?

- how do I route the service drops in the hallway best to minimize the size and extend of the molding?

The number of questions that you have to ask yourself may be endless. You may not be able to come up with all the answers. But, whatever you can determine will save you money and eliminate a lot of aggravation later. To determine where you desire the equipment to be installed is better than finding out that the contractor has put it exactly where it should not be.The contractor will be happy to move it, as long as he gets paid again.

At this point in time, all you have to do is to record what you have seen and describe what you think should be done in the best interest of your enterprise. A few notations and sketches that can be included in your bid documentation are of great importance to illustrate how you expect the work to be done and will give you peace of mind during the implementation phase.

Murphy's law applies to HFC broadband networks as well.

Chapter 12

Acceptance-Testing and Documentation

Testing procedures during implementation and acceptance-testing after the completion of the system are important activities to establish the integrity and quality of the installation.

To be in the position to better monitor the activities of the contractor, it is recommended to establish the testing program in conjunction with the milestones of the implementation schedule.

It is a good idea to subdivide the HFC system under construction into segments so that each have their own completion dates and test activities. For instance, it is possible to take a single fiber-optic cable branch and the coaxial inside wiring in one building and make it a separate segment. Each of these segments then carry their own milestones for the completion of installation activation and testing, which can then be more closely monitored by the project supervisor.

This method of segmentation makes the project more measurable, requires less program management on the part of the owner and safeguards against delays. Without segmentation and milestone completion dates for each segment, the contractor will complete the entire installation first before any testing is started.

Admission Tests

Cable Reel Testing

Both single-mode fiber-optic cables and coaxial cables shall be tested by the contractor upon receipt of the cable reels.

Fiber-optic Cable OTDR Admission Test

The Optical Time Domain Reflectometer (OTDR) is used to measure the cable attenuation of the fiber strands. The OTDR is a single-ended device which transmits a short light pulse and analyzes the reflected pulse. A printed trace result with date shall be provided by the contractor before commencement of any cable installation.

The contractor shall make a deployment plan for all fiber cables so that the OTDR testing data of the fiber reel can be later compared with OTDR tests after completion of the installation. Reels showing any discontinuities shall be rejected and replaced.

Coaxial Cable TDR Admission Test

The Time Domain Reflectometer (TDR) is used to measure the length of the cable on the reel as well as determine any damages, kinks or reflections. The TDR is a single-ended device which transmits RF energy through the cable and analyses the reflected energy. The cable end can be either terminated, shorted or left open. A printed trace result with date shall be provided by the contractor before commencement of installation.

The contractor shall make a deployment plan for all coaxial cables so that the TDR testing data can be compared with TDR testing after cable installation. Reels showing discontinuities of any kind shall be rejected and replaced.

While in storage, all larger coaxial cable reels must be stored resting on the reel flanges and not be stacked.

Passive Equipment

Passive equipment such as splitters, directional couplers shall be spot-checked for performance. The sample rate shall be 10%. The test can be conducted using a noise generator producing white noise across the frequency band and a scope display to view and print the performance over frequency. Alternatively, a sweep generator can be used. The test data shall be submitted to the owner.

Amplifier Burn-in

Amplifiers shall be bench-tested and burned in for a period of 24 hours before deployment. Carrier frequencies may be produced by RF modulators set for the highest and lowest frequencies of the passband. A sweep generator is used to check the frequency response of the amplifier. The combined carrier and swept response picture is displayed on the scope of the sweep receiver. Print-outs are provided for the trace at the beginning and the end of the 24-hour test period.

Fiber-optic Transceivers

Fiber-optic transmitters and receivers are checked in the same manner than coaxial amplifiers.

After interconnecting the transmitter and receiver using a IO dB optical attenuator, the input of the fiber-optic transmitter is connected to the modulators and the sweep generator representing the entire passband. The fiber-optic receiver is then connected to the sweep receiver for the recording of RF levels and flatness.

Print-outs are provided, again, at the beginning and at the end of the 24-hour period displaying the overall performance over the frequency band through the optical path of the transmitter and receiver.

Functional Testing

Functional testing is performed by the contractor during construction of the system. These in-progress tests can be performed for each plant segment separately.

Completed functional testing confirms that the performance of each system segment meets the design criteria.

Fiber-Optic Cable OTDR Testing

Upon completion of all cable pulling and installation of a cable section, the fiber-optic cable shall be tested with the OTDR. This test is to be conducted before commencement of splicing to confirm the integrity of the cable after it has been subjected to the stresses of the installation.

A comparison shall be made of the printed traces with the traces formerly obtained during the admission tests. The test data shall be submitted to the owner.

Coaxial Cable TDR Testing

All coaxial cable segments shall be tested, using the TDR method, after completion of the installation and before splicing or connectorization. The length of all cable sections must be recorded together with the TDR print-out. The data will be submitted to the owner and compared with previously taken cable-reel tests. The distance information shall be compared with the distances that have been used in the design of the system.

Major differences (over 20 ft.) shall be noted and used to update the system design data.

Physical Inspection

In addition to the TDR testing, the contractor shall conduct a physical inspection jointly with the owner.

All cables and cable ends shall be inspected to locate deformed cable ends, heavy damage to the jacket, kinks, dents, flattened sections or other abrasions. Any cable run with such impairments may be rejected and may require replacement before commencement of splicing.

Activation and Sweep-Testing
in the Forward Direction

Upon activation and level-setting of the coaxial amplifiers, the optical transmitter and fiber-node receiver, the HFC segment shall be sweep tested in the forward direction.

The test setup consists of modulators at lowest, middle and highest passband frequencies with the sweep generator connected to the RF input of the fiber-optic transmitter at specified input levels. In case of a sub-split system, the modulators will be operating at 54, 220, 550 and 750 MHz. The same frequencies, except for 54 MHz, can be used for high-split system testing.

The sweep receiver shall be connected first to the fiber-node receiver to observe levels and flatness of the RF output. The sweep receiver is then moved to the next and any following coaxial amplifier. Level and slope adjustments as well as flatness corrections shall be made before making a printed record.

The swept-response pictures will be reviewed by the owner and become a part of the documentation package. The swept-response pictures indicate the flatness and overall frequency response of the HFC transmission system. Since the sweep generator rapidly changes the transmitted frequencies across the passband, any frequency response problems in the fiber-optic and electrical segment will be noticed.

The peak-to-valley response in any 6 MHz channel shall be within ±1 dBmV. The peak-to-valley response across the frequency band shall be within ±3.0 dBmV. Major deviations from these specifications point to installation problems.

Sweep-testing shall be repeated for every coaxial cable branch of each HFC segment and compared for compatible results. Any sudden change of the frequency response or any sinusoidal structure of the response indicates bad splicing work, loose or insufficiently tightened seizure screws or too long connector pins. Immediate cure of such conditions is a high priority before continuing the testing program.

Activation and Sweep-Testing in the Return Direction

Since the HFC system has a tree-and-branch segment, there are more than one system end. In order to test every branch in the return direction, the sweep transmitter and modulators for 5 and 42 MHz sub-split systems or for 5 and 186 MHz high-split systems must be moved to the last amplifier in each branch.

A +17.0 dBmV is required at the input of any reverse distribution amplifier. The same level must be measured at the input to the amplifier feeding the fiber-optic transmitter.

After adjusting the pad values at the output of the return amplifier in each branch to obtain the +17.0 dBmV input, the amplifier feeding the optical transmitter can be set up to assure the correct transmission through the optical segment.

Swept response pictures will be taken at the RF output of the fiber-optic receiver at the control center. The response traces will be reviewed by the owner and become a part of the documentation package.

The peak-to-valley response in any 6 MHz channel shall, again, be within ± 1 dBmV. Because of the narrow return passband, the peak-to-valley performance across the frequency band shall be within ±2.0 dBmV.

Outlet-Level Testing

One-hundred percent of all outlets must be tested to confirm system design and to assure the integrity of the system. There are two permissible options to perform outlet-level testing.

> a) Connect the signal-level meter to each outlet F-connector and record the levels of the specified carriers.
> (Sub-split - 54, 220, 550 and 750 MHz)
> (High-split - 220, 550 and 750 MHz)
> Record all levels by drop numbers and compile for documentation submittal

or, since all service drops are of equal length, the level measurement may be taken at the multitap ports at the IDF, provided that each drop has been tested for continuity.

> b) During connectorization of the service drop, perform a continuity check. The drop cable must be connected to the outlet terminal, which will be terminated with a 75-ohm F-terminator. The measurement is made using an ohm-meter at the IDF end of the drop between the outer shield and center conductor of the completed F-connector. Continuity checking is recorded by service drop numbers.
>
> After completion of activation and sweep-testing, the levels of the specified carriers are measured at the multitap ports at the IDF. Each multiport is tested and associated with service drop cable, which will be connected when removing the signal level meter from the port. All measurements are recorded by multiport number, multiport level and service drop number.

Outlet-level test records will be submitted to the owner for review and approval and become a part of the documentation package.

Cumulative Leakage Index (CLI) Testing

Cumulative leakage tests are only required for coaxial segments of the HFC system.

CLI tests can only be conducted after completion of activation and interconnection of all service drops. Any segment of the network can be tested separately.

The CLI test is an important indicator of the quality of splicing and of the integrity of system construction. This most important test is subject to FCC and FAA regulations. The test shall be conducted in accordance with FCC specifications FCC 76-601.

This test is extremely important to verify good signal transmission capabilities. Measurement of RF leakage from a cable section will verify RF shielding integrity of the system. Sections of cable with low signal leakage will not allow transmitted signals to escape from the system. Unwanted noise and extraneous signals can also infiltrate the system, which cause unacceptable system performance.

When an RF leak of a significant level is found, repairs should be made immediately. Proper connectorization of the coaxial cable and special care taken in the interconnection of devices will assure a leak-proof or "tight" system.

CLI testing can be conducted with portable scanners or calibrated dipole antennas. It is important that every portion of the coaxial system be checked between the fiber-node location and every outlet by walking the system in a distance of 6-IO feet from the cable.

CLI testing shall be certified by the contractor by describing the segment that has been tested and that leakage has not been measurable.

Acceptance-Testing

Acceptance-testing cannot be performed until all HFC plant segments have been completed and all functional tests have been performed. Your presence during the acceptance-testing phase is mandatory, as it is the contractor's demonstration of the completion of the project.

At the time of the acceptance-testing, it is recommended that all records from functional testing, previously completed, are available for review. This should include:

- amplifier alignment data with pad and equalizer values
- CLI certifications
- swept response pictures for forward and return segments
- outlet levels

Acceptance-testing consists only of end-to-end tests that will prove the quality of performance of the entire HFC network.

Carrier-to-Noise Ratio (C/N)

The carrier-to-noise test is the most important indicator of system performance. The test shall be performed for each HFC segment. If, for instance, the HFC system consists of 5 fiber-nodes with coaxial distribution and I coaxial-node, the C/N ratio measurement shall be taken at the extremeties of all system segments, i. e. at the test tap of the last amplifier.

The C/N ratio is determined by reading the carrier levels with a signal-level meter of high quality and subtracting the noise-level reading from the carrier level. To read the noise floor, move the signal-level meter to spaces between carriers and optimize for the lowest reading. The S/N switch on the meter must be switched to permit a 4 MHz bandwidth range.

Record both carrier level and noise readings. The algebraic difference is the C/N ratio, which can be compared to the design calculations. The C/N measurement shall be performed for all carriers in both sub-split and high-split systems, i.e. 54, 220, 550 and 750 MHz (sub-split) and 220, 550, 750 MHz (high-split).

All C/N ratio measurements shall become a part of the documentation package.

The Hum Component

Hum is introduced into fiber-optic and coaxial transmission equipment by coupling or induction from the IIO Vac power network. The hum test should be conducted at the same end points of the system as the carrier-to-noise measurements, i.e. at an end-of-line multitap.

The hum test is conducted by tuning the signal-level meter to each of the carrier frequencies and using the hum-test switch. The measurement is made as a percentage of the signal level. The hum level shall be less than 2%. Record all measurements for inclusion into the documentation package.

Loop-Testing

Loop-testing assumes that the control center of the HFC system is equipped with RF demodulators and modulators or an RF translator unit is available to loop signals arriving on the return system to retransmission in the forward direction.

Loop-testing will establish the uniformity of each of the HFC segments and shall be performed at the end-of-line locations that are being used for C/N and hum-testing.

At an outlet location, connect a camera, a VCR or an NTSC card equipped computer to an RF modulator or use a T-I data stream into an RF T-I modulator. Use the highest frequency assignment for the return transmission. Using a signal-level meter, measure the loop signal level at an adjacent receive outlet. Record the receive-level measurement and move the signal-level meter to all other HFC end-of-segment locations to repeat the measurement.

All receive levels shall be within 3 dB of each other. If they show wider variations, it is prudent to check amplifier alignment and fiber-link setting to reduce the loop level difference.

The test may be repeated using one transmit point at the end of each HFC segment and taking level readings at all other HFC segment end points. Test results become a part of the documentation package.

Operational Tests

The scope of any operational testing is directly related to the equipment that will be initially deployed in the control center. If you are just implementing the HFC infrastructure under the initial procurement, no operational testing is required.

If, on the other hand, you are including matrix switches, recorders, editing equipment, video retrieval, distance learning or T-I interfaces to the outside world, then it is a good practice to set up operational test procedures.

Especially when you have included automation software for programmable operation of switches, channel selection, recording and play-back functions, it is important to incorporate the requirement for the demonstration of every computer-controlled function in the scope of the operational testing program. The request for training may also be included. Automation software varies widely relative to functionality, capabilities and complexity.

In addition to automation software, there is HFC network-management software that can monitor the performance of any component in the network, when equipped with special sensing equipment. These network-management systems can aid in the prevention of outages, locate outages, provide graphic displays of the network and operate redundancy switches for alternate fiber routing. It is obvious that this added degree of complexity requires operational testing and training before it can be accepted.

Documentation

The documentation of both installation activities and test data is often called deliverables or contract data requirements. Your RFP or IFB should include, as a minimum, a listing and a brief description of the documentation that you expect to receive from the contractor upon completion of all installation and testing.

Installation Documentation

A comprehensive list of installation particulars may contain the following requirements:

- a CAD drawing showing the outside-plant routing of the HFC network with conduit and cable numbers for both fiber and coaxial sections

- a CAD drawing showing the electrical design of the HFC system with all components in place indicating the design levels of all branches inclusive of amplifier setup data and multiport levels

- a tree diagram showing the in-building installation with all components inclusive of amplifier setup data and actual multiport levels, as tested

- a listing of all cable numbers with identification of origin and destination by HFC segment

- a listing of all service drops with cable numbers and identification of origin and destination by building and segment

- a CAD presentation of all MDF and IDF locations depicting the equipment as mounted on the wallboard with cable numbers, origins and destinations of all cable

- a comprehensive CAD rack-up and rack-wiring presentation of any headend and control center equipment with input and output levels of all equipment

Test Documentation

The test documentation package should include all test data that was compiled during the testing phase of the project such as

- OTDR traces of fiber-optic cable reel admission testing

- TDR traces of coaxial cable reel admission testing with cable lengths and reel numbers

- OTDR traces of fiber-optic cables after installation with cable number and crossreference to the reel number

- TDR traces of all coaxial cable sections with distances, cable numbers and information on location, origin and destination

- setup data of fiber-optic transmitters and fiber-node receivers for every forward and return fiber link

- setup data of coaxial amplifiers with pad and equalizer values for both forward and return operation

- multiport or outlet levels of every service drop inclusive of cable number, outlet location and multiport port number

- swept response pictures of all fiber and coaxial branches in the forward directions and as taken at the end-of-line locations

- swept response pictures of all coaxial and fiber branches in the return direction and as taken at the headend

- records of CLI tests for all coaxial branches

- acceptance-test records of C/N ratios of all branches

- acceptance-test records of hum measurements at all end-of-line locations of all branches

- documentation of all loop tests conducted with identification of transmit, receive location and receive levels measured

- sketches showing the test equipment and other equipment setup and levels used for every test setup used during the testing program

Miscellaneous Documentation

It is good practice to request commercial grade manuals for every equipment that has been provided by the contractor. In addition, you may ask the contractor to secure specific maintenance recommendations and trouble-shooting guides from the vendors.

Contract Data Delivery

The approval by the owner of the acceptance test data and the acceptance of all documentation and deliverables that have been requested in the RFP or IFB usually indicates that all contract activities are completed. This event is also the begin date of the contractor's warranty. It is a good practice to have the contractor extend all manufacturers' product warranties for the full duration of the agreed upon system warranty period.

Chapter 13

The HFC Proposal Specifications

You are probably quite familiar with competitive bidding practices, whether they are RFPs or IFBs. You might even have considerable experience in dealing with contractors, evaluating proposals, writing contract terms and conditions and getting the project completed on time. It is not the purpose of this book to repeat what you already know, but to emphasize areas of importance as they relate to the implementation of HFC networks.

The HFC broadband network proposal specifications should be developed in great detail and identify

      a)   what needs to be done
      b)   how is it to be done
      c)   who is the responsible party

Any ambiguities in the proposal specification will lead to misinterpretations that may cause substantial differences in costing and may also produce reasons for delays and extra scope requests during the contract period.

Qualifying the Bidders

Before issuance of your bid request, you will probably investigate some of the regional implementation contractors. Ask for references of completed HFC systems. Chances are that not many telecommunication contractors have been exposed to the construction of end-to-end HFC systems. And if they have, their experience is limited to outside-plant installations with both fiber-optic and coaxial cables on poles.

Another category of contractors are the satellite service and apartment-house wiring companies. They may give you numerous references of satisfied customers, but do not have the expertise required for the complexity of an HFC system. What you are looking for is the combination of an UG contractor, a fiber-optic specialist and a coaxial-cable installation expert.

So, take time out to investigate the experience of some of the companies in your area. The local cable companies may give you some leads, or may be interested themselves. The local exchange company, or operating telephone company, probably can advise you of network service companies specializing in implementation services.

Besides references of past work, it is important to field their expertise. A good preselection test is the request for a list of test equipment that would be used on the project requiring a 750 MHz passband. An existing test equipment complement is a good indicator of competency. If TDRs, OTDRs, sweep-test equipment, spectrum analyzers and signal-level meters are available, then there is a good chance that the company can comply with your specifications and deliver a first-class installation.

But, assuming that you have successfully preselected a number of interested bidders, we will concentrate on some of the details of the proposal specifications in the following.

Scope of Work

The scope-of-work section of the proposal specifications should provide an overview of the project as well as an emphasis of your goals and expectations.

System Description

The system description should state the purpose of the project and describe the planned HFC architecture and capacity. Detailed requirements for outside plant and in-building cabling should be included to emphasize the quality requirements. You may want to include routing diagrams and sketches of the proposed system. Segment the system implementation into geographical areas consisting of fiber-optic and coaxial components of similar size and develop a schedule of installation priorities.

If you offer your design drawings and Bill-of-Material as the basis of the HFC architecture, do not include any design calculations and make the contractor responsible for design verification and Bill-of-Material omissions.

You may want to make a list of all applicable specifications that are considered an integral part of the proposal specification.

In summary, write your proposal specifications as if you are writing an implementation contract so that, when the winning bidder has been selected, only minor work is required to finalize the contract document.

Responsibilities of the Contractor

Explain the responsibilities of the contractor in some detail. If you are procuring some of the hardware, describe in detail the type and quantity of what you will provide.

Make the contractor responsible for the

- procurement of any and all hardware equipment

- engineering of all installation activities in accordance with the provided installation specifications

- verification and acceptance of the owner-provided design as well as total responsibility for the electrical and mechanical design of the system and the meeting of all stated performance specifications.

- performance of any remedial activities resulting from errors made in the electrical design of the system

- verification of the Bill-of-Material provided by the owner and provision of any equipment or hardware that may have been purposely or inadvertently omitted

- installation of the system and components in accordance with the provided installation specifications and routing diagram

- testing and acceptance-testing of the HFC system in accordance with the test specifications

- delivery of a functioning system meeting the performance specifications of the test specifications and the design documentation

- delivery of contract data deliverable requirements in accordance with the documentation section of the test specifications

- immediate repair and remedial action to any damages done to the owner's property

- timely completion of all activities within the scheduling requirements of the project.

The Project Timetable

You may have a deadline in your mind for the completion of the project. Give yourself some time to complete the proposal specifications and give the vendors a four-week period to respond. Encourage their participation in a site-visitation or even make it mandatory for any prospective bidder to attend. An HFC system project cannot and should not be bid blindfolded.

Establish a question-and-answer period of one week after the site visitation and exchange all questions and answers with all prospective bidders, and in writing.

Develop a project time-line that you want to see met by the winning contractor. Recognize that the contractor requires time for the procurement and on-site delivery of cable and equipment. Subdivide your HFC network into controllable segments and indicate your preference of the sequence for completion of each individual segment.

Detail each segment and develop milestones for each of the major installation and test activities, i.e.

- admission testing
- cable-pulling OSP
- cable installation ISP
- equipment installation and connectorization
- activation and functional testing
- physical inspection
- review and approval of functional test data

When the last segment is activated and all functional tests have been performed, the acceptance-test period can be scheduled for the entire network.

Submit this timetable in the proposal specification package and request concurrence and/or permit the bidder to present an alternative implementation schedule.

If it is standard practice by your company to make progress payments on a percentage-completion basis, you may want to determine completion percentages for each segment and base the percentage on the complexity of each segment.

Since cash flow is important to every contractor, you may want to develop incentives for early completion and a sliding scale of deducts for segments not completed on time.

Bid-Response Requirements

Whether you are requesting the bids in an IFB or RFP format, it is important that you specify what information you require to be provided in the bid response. A sample of important technical bid-response requirements is listed in the following (general terms and conditions are not included):

- An equipment list of all system components to be supplied with model number, manufacturer's name and compliance to the proposal specifications

- A point-to-point response to the technical specifications

- A point-to-point response to the scope of work

- Acceptance of the owner-provided system design or the submittal of a preliminary design by the bidder, which is explicit enough to identify the bidder's expertise with HFC system design

- Acceptance and comments on project scheduling

- A staffing plan

- Assignment of an on-site representative

- References of completed HFC systems and names of contacts

- A list of test equipment that will be made available for the project

- A list and description of commonly used project progress control methods

Your selection process may be simplified if one of the lower bidders provides detailed and believable responses to your request.

Owner-provided Activities

The proposal specification should identify any work or equipment that you will provide for the project.

Provision for IIO Vac Power

The provision of electrical power is usually one of the responsibilities of the owner. On your visits to MDF locations, you may have already noted that new electrical circuits are required for optic-node receivers and coaxial-cable power supplies.

In a large enterprise the resident electrician can provide the feed. It is a good practice to develop a new or extend your present UPS system to these locations. Only fiber-node receiver locations require IIO Vac service. The headend or control center may be the one location where multiple phases and circuits are required.

The proposal specifications should mention your intention to provide electrical power.

Storage and Office Space

The larger the project, the more cable and equipment will have to be stored. As an accommodation to the contractor, the on-site availability of a storage area eases the coordination problems.

Large projects may require outside storage for cable, inside storage for equipment and bench-testing as well as a small project office. Small and single-building projects, of course, cannot afford such frills except for a desk in a partially occupied office area.

Whatever can be provided will be appreciated by the contractor and your intentions should be mentioned in the proposal specifications for the bid invitation.

Access

Accessibility to the places of work is a main cause for delays in the completion of any project. In an active building, there is never a good time to string cables. Telephone closets, in most cases, do not open with a masterkey. Every closet may have a different lock or every building has a different set of keys. Escorting and scheduling of some work to after the closing of the regular workday may relieve some of the problems.

Whatever you can offer relative to the improvement of accessibility, or whatever restrictions of working hours you may be contemplating, explain your approach in the proposal specification. A word of caution - unlimited accessibility during the hours of 8:00 am to 5:00 pm assures lowest costs.

Project Management Requirements

Proven project management techniques are often applied to implementation contracts to assess productivity, work in progress and percentage of completion of any work phase. Project management techniques can be taken to extremes by imposing work-breakdown structures, pert charts and critical path analysis.

A simple procedure, however, can be used to control progress and completion percentages. Divide the project into segments, list the work activities of each segment in the sequence of occurrence, assign time periods with completion dates to every activity and monitor the progress of the work by comparing completed milestones with the time periods of the plan.

It is recommended to request a detailed schedule with milestones for major activities from the bidder as a part of his proposal. The response will permit you to assess the ability of the contractor to control the project activities.

Besides this request in the proposal specification, you may want to include your own coordination requirements, i.e.

- the reporting of work performed with completion percentage

- the frequency and scheduling of progress meetings

- the content of weekly progress reports

- the reporting of problems encountered or difficulties experienced

- the handling of changes and additions

- the implementation of quality-control measures

If you can define clearly in your proposal request what reports have to be provided, when to provide them and what details you expect, then there will be no surprises at a later date. It will also reduce your workload if you do not have to seek the whereabouts of your contractor's representative to find out what has been accomplished.

Technical Specifications

The technical specifications of cables and equipment that you expect to be provided with should be added to your proposal specifications as an attachment.

Hardware and Equipment

These specifications may be similar to the equipment specifications provided in Chapter 8. Since vendor names and model numbers are not given, you may ask the bidder to respond with an identification of manufacturers, model numbers and reasons for the selection of the product.

If, on the other hand, you have already checked and compared the products of some of the manufacturers and you are convinced that a particular make is better suited for your system, make it known to the bidder that you are looking for a specific brand.

It is good practice to request samples of some parts and equipment from the winning contractor so that you know what you are getting and can make informed decisions before the commencement of the implementation activities.

Alternatively, you may want to find out if there are any substantial savings if you would procure major hardware items and equipment directly. Prices and discounts are easily obtained from the major distributors that supply cable TV and telephone companies. In cases where you already know what the prices are, you may ask the bidder to propose the project as a full turnkey and as a modified turnkey, which would exclude the items that you will provide.

This alternative is harder to manage as you become responsible to have the equipment available on time. Split procurements lead to split responsibilities and can create coordination problems. If, however, the cost savings are substantial, the additional burden may well be justified.

Software

When dealing with automation, network management or video retrieval software, no education can be too thorough. Whatever the winning bidder has proposed should be investigated in detail, even if it means a visitation to the vendor for a demonstration.

In the alternative, you may request the bidders to provide more than one solution in their proposals with the understanding that the final decision will be made after contract award and after thorough evaluation between the owner, the contractor and the vendor.

Installation Specifications

Your installation planning, which you have compiled in accordance with Chapter II of this book, is all you need. The installation standards that you want to have observed, and that you worked so diligently to compile, must become a part of your request for a proposal. Again, the information package could be issued as an attachment or appendix with a simple statement of these groundrules. The bidders will either comply or take exceptions to a few things, but, in any event, you have some assurance that your proposed HFC system will have an orderly appearance.

Besides including the installation standards of your universal wiring plan, make sure to also include your outside-plant and inside-plant routing sketches so that the bidder better understands what your requirements are. Make the contractor responsible, however, for the provision of sketches of proposed installation particulars, for your approval, prior to the commencement of any work.

In the alternative, only submit your universal wiring plan with installation standards and request that the winning contractor's start of construction is subject to a field investigation of the entire cable route and agreement on conduit, cable routing and equipment-mounting details. This way, it would become the responsibility of the contractor to make sketches of critical locations and seek your approval before proceeding with the work.

While this alternative approach may cause a delay at the beginning of the project, it is a good approach and will assure that the contractor understands your requirements without you having to expose your drafting skills. Any misinterpretation of your drawings will also be avoided.

Acceptance-Test Specifications

This attachment to the request for proposal should comprise all test specifications of Chapter I2. This should include admissions tests, functional tests and acceptance-testing. Physical inspections are also considered a part of the scope of testing and you may want to alert the bidder that periodic physical inspections of completed plant portions will be scheduled by the owner and shall be conducted jointly with the contractor's resident engineer.

Considering that you are building an HFC system with an upper frequency of 750 MHz, there will not be many contractors that own test equipment capable of testing these frequencies. Obviously, this will not be a problem a year or two from now when all the cable companies complete their HFC projects. But at this point in time, it is important that you find a contractor who can perform these specific tests.

The minimum test-equipment complement should consist of

- OTDR - Optical Time Domain Reflectometer
- TDR - Time Domain Reflectometer
- Sweep Transmitter (5 to 750 MHz)
- Sweep Receiver (5 to 750 MHz)
- White Noise Generator to 750 MHz
- Signal Level Meter (5 to 750 MHz)
- Optical Power Meter (1310 nm)
- Assortment of Optical Attenuators
- Assortment of Coaxial Attenuators
- Accessory Cables and Connectors - as required

Your request for proposal should require the bidder to provide information on presently owned test equipment and test equipment that will be available for contractual testing. Since contractors have the tendency to move expensive test equipment between job sites, it is recommended to make the bidder and winning contractor aware of the fact that all test equipment must stay on the job site until project completion.

Documentation Requirements

The contract data requirements of a typical HFC system have been listed in considerable detail in Chapter 12. You may want to give these requirements your personal touch and add your specific requirements to both the installation documentation and the test documentation that has been listed in this Chapter. In cases where your company has an established Facilities Management System, your list may require FMS details to be included.

Since compliance by the bidder to your listing of documentation items does not assure you a usable document, you may want to state in the request for proposal that the exact format and content of the data deliverables will be determined during the contract implementation phase and should be provided by the contractor for the approval by the owner.

Thoughts in Closing

You are now at the point of requesting a bid for the implementation of your own HFC enterprisewide broadband network. Since it can handle any future voice and data network requirements via cable modems, it may be the last infrastructure that you ever have to build. The outside world sure seems to think so. Cable television and telephone companies would not build an architecture that may become obsolete in a 20-year period.

HFC has become a favored long-term broadband access platform because it combines economical investment with flexibility of transmission, highest reliability and quality of performance. As the transmission medium for the next century, the HFC network is a medium which is capable of delivering heterogeneous traffic of various bandwidths, modulation formats and latency characteristics in circuit-based, channel-based or packet-based formats.

Your efforts of following through with the design of the network, the development of installation standards and test specifications are commendable. You have developed the broadband infrastructure of the future. You are a big step ahead of your competition.

The HFC broadband infrastructure can be built at a fraction of the funding that you would need for current high-speed data networks on copper, or even multimode fiber. Your capital investment is scalable with expenses for terminal equipment deferred until your user-demand emerges and until inexpensive cable modems make your HFC system the backbone facility for any future expansion of your communication needs.

What is left to do is the selection of a capable contractor to assure your company the highest reliability and quality of performance that the HFC broadband network has to offer.

With virtually no costs required for upkeep, repair and maintenance, your HFC network will satisfy your increasing voice, data and video demands for decades. The HFC system will connect your enterprise to the information superhighway using the identical standards of the serving local exchange carriers of the future, i.e. fiber for distance, coaxial cables for distribution and ATM switches connected to SONET rings for the transportation of signals of any speed and bandwidth.

INDEX

Other Great Books We Publish

Flatiron Publishing publishes books and magazines and organizes trade conferences on computer telephony, telecommunications, networking and voice processing. It also distributes the books of other publishers, making it the "central source" for all the above materials. Call or write for your FREE catalog.

ATM Users' Guide
City, County State Guide To Acquiring A Phone System
Client Server Computer Telephony
Computer Based Fax Processing
Customer Service Over the Phone
Frames, Packets and Cells in Broadband Networking
The Guide to Frame Relay
The Guide to SONET
The Guide to T-1 Networking
Local & Long Distance Telephone Billing Practices
Moore's Imaging Dictionary
Newton's Telecom Dictionary
PC-Based Voice Processing
SCSA — The Definitive Reference Manual
Speech Recognition
Telephony for Computer Professionals
Traffic Engineering Handbook
VideoConferencing: The Whole Picture
Which Phone System Should I Buy?

Quantity Purchases

If you wish to purchase this book, or any others, in quantity, please contact:

Christine Kern, *Manager*
Flatiron Publishing, Inc.
12 West 21 Street New York, NY 10010
212-691-8215 1-800-LIBRARY
facsimile orders: 212-691-1191
Internet harrynewton@mcimail.com

Printed and bound by CPI Group (UK) Ltd, Croydon, CR0 4YY

21/10/2024

01777098-0010